Frictionlessness

thinking|media

series editors:

bernd herzogenrath
patricia pisters

Frictionlessness

The Silicon Valley Philosophy of Seamless Technology and the Aesthetic Value of Imperfection

Jakko Kemper

BLOOMSBURY ACADEMIC
NEW YORK · LONDON · OXFORD · NEW DELHI · SYDNEY

BLOOMSBURY ACADEMIC
Bloomsbury Publishing Inc, 1385 Broadway, New York, NY 10018, USA
Bloomsbury Publishing Plc, 50 Bedford Square, London, WC1B 3DP, UK
Bloomsbury Publishing Ireland, 29 Earlsfort Terrace, Dublin 2, D02 AY28, Ireland

BLOOMSBURY, BLOOMSBURY ACADEMIC and the Diana logo are trademarks
of Bloomsbury Publishing Plc

First published in the United States of America 2024
Paperback edition published 2025

For legal purposes the Acknowledgments on p. vii constitute an extension
of this copyright page.

Cover design: Daniel Benneworth-Gray
Cover image © Paolo Sanfilippo

Bloomsbury Publishing Inc does not have any control over, or responsibility
for, any third-party websites referred to or in this book. All internet addresses
given in this book were correct at the time of going to press. The author and
publisher regret any inconvenience caused if addresses have changed or sites
have ceased to exist but can accept no responsibility for any such changes.

A catalog record for this book is available from the Library of Congress.

Library of Congress Cataloging-in-Publication Data

Names: Kemper, Jakko, author.
Title: Frictionlessness : the Silicon Valley philosophy of seamless
technology and the aesthetic value of imperfection / Jakko Kemper.
Description: New York : Bloomsbury Academic, 2024. |
Series: Thinking/media | Includes bibliographical references and index.
Identifiers: LCCN 2023025568 (print) | LCCN 2023025569 (ebook) |
ISBN 9798765104415 (hardback) | ISBN 9798765104422 (paperback) |
ISBN 9798765104453 (epub) | ISBN 9798765104446 (pdf) |
ISBN 9798765104439 (ebook other)
Subjects: LCSH: Software failures–Philosophy. | Digital media–Philosophy. |
Reliability (Engineering)–Philosophy. | Imperfection–Philosophy.
Classification: LCC QA76.76.F34 K46 2024 (print) |
LCC QA76.76.F34 (ebook) | DDC 004–dc23/eng/20230601
LC record available at https://lccn.loc.gov/2023025568
LC ebook record available at https://lccn.loc.gov/2023025569

ISBN: HB: 979-8-7651-0441-5
      PB: 979-8-7651-0442-2
    ePDF: 979-8-7651-0444-6
   eBook: 979-8-7651-0445-3

Typeset by Deanta Global Publishing Services, Chennai, India

For product safety related questions contact productsafety@bloomsbury.com.

To find out more about our authors and books visit www.bloomsbury.com
and sign up for our newsletters.

Contents

Figures

Acknowledgments

I like to think that this project began in 2015, when, on an overcast day in Oslo, I came across a quote by John Peel printed onto the window of a record store. The quote stuck with me, kindled my interest in the affective quality of technological imperfections, and now furnishes this book's introduction. In the time between that chance encounter in Oslo and the present, this project thus came into being—slowly, haphazardly, sometimes maddeningly, and with the indispensable help, care, and support of the following people.

I am first of all grateful to Patricia Pisters and Bernd Herzogenrath for believing in this project and giving it a home in Bloomsbury's Thinking Media series. This book has also benefitted from valuable comments from and conversations with Marie-Aude Baronian, Carolyn Birdsall, Steven J. Jackson, Caleb Kelly, Esther Peeren, Ellen Rutten, Yuriko Saito, and Niels van Doorn.

I was lucky enough to write parts of this book as a researcher at the Amsterdam School for Cultural Analysis (ASCA). For every story about research as a lonely and stressful undertaking, ASCA shows that there is also plenty of laughter, comradery, and collaboration to be found. Thank you, Eloe, Esther, Patricia, Jaap, and Jantine, for your tireless work in making ASCA such a hospitable and invigorating academic home. Thank you, moreover, to ASCA's vibrant community of researchers, for all the support offered, drinks shared, conversations had, and reading groups organized. Fabienne Rachmadiev deserves special recognition here for being my fellow traveler in the imperfect. Special thanks also go to Alex Williams, who steadfastly offered an exacting eye and good spirits throughout this book's development. His critical commitment to my work and our many conversations on vaporwave, spectrality, and the finer points of French theory have been indispensable, especially to the first and fifth chapter. Thank you!

I am also grateful to Steven J. Jackson, who generously hosted me as a guest researcher at Cornell University and whose invaluable comments on various drafts pushed my thinking in crucial new directions. The first two chapters and the Coda especially benefited from his insightful feedback. Many thanks also to the people in Cornell's Department of Information Science, who did so much to make me feel welcome. My time at Cornell has been one of the highlights of this research project and the added privilege of residing in Ithaca precisely during the transition from summer to autumn helped set the affective tone for much of my thinking on the appeal of decay and disintegration.

The latter part of this book was completed during my time as a lecturer at the University of Amsterdam's Department of Media Studies. I could not have wished for a more nourishing environment and for more supportive colleagues. It is unfortunate that the book was already nearly done when I started to work at the department; in another world, I would have met my colleagues during the more formative stages of my research, and the book would have been that much stronger for it. Nonetheless, these colleagues provided crucial support during the final stretches of writing. I feel lucky to be surrounded by so many brilliant people—here's to many future collaborations.

I thank the artists whose work I analyze for allowing me to reproduce images of their work in this book. Jornt Elzinga deserves special mention here, as his records have soundtracked large parts of the writing process. Recognition also goes to the excellent editorial team at Bloomsbury and, especially, to Katie Gallof, who has been a joy to work with.

I am infinitely grateful to my friends outside academia. Being able to occasionally leave the confines of my research behind helped tremendously in maintaining (some of) my sanity throughout these years. You have helped more than you all know. Lastly, I want to thank my parents, Lex and Ria. While I opened these acknowledgments with the suggestion that this project began with a fortuitous encounter in Oslo, the truth of the matter is, of course, that it started long before that. It started with being taught to regard the world with a sense of curiosity and compassion, with being raised in a house of music and books, and with the generously imparted knowledge that there is always the bedrock of family. For this and so much more, thank you.

Introduction

Of the many anecdotes that the late John Peel (1939–2004)—British radio presenter and disc jockey—left in his wake, there is one that epitomizes the stakes of the present study. This anecdote describes Peel being confronted by an individual who argued for the superiority of the compact disc (CD) over vinyl because the former medium lacks the latter's distinctive surface noise. Faced with this claim, Peel pithily riposted: "Listen, mate, *life* has surface noise" (Hussey 2016, emphasis in original). Peel, as music journalist Allison Hussey (2016) underlines, here "defends vinyl's imperfections" by ascribing a profound value to flawed mediation. Vinyl's faltering qualities, so Peel implies, offer an authentic reflection of the vicissitudes of life. By contrast, the noiselessness of the compact disc, posing a frictionless alternative to vinyl's crackling miscellany of hums and hisses, is framed as affectively incongruous with the messy reality of existence. In this regard, Peel's affection for vinyl can be seen as indicative of a wider human desire to find, within the welter of available technologies, something that resonates in pleasing harmony with lived experience. Peel's words suggest, moreover, an emotional attachment to technology itself, a pronounced investment in its fallible and finite materiality.

This affirmation of vinyl's ostensible flaws connotes an affinity for *a technological aesthetic of imperfection*, or, phrased differently, an aesthetic of imperfection as it is encountered in technological or technologically produced objects and processes. Peel's account—which, as I will shortly discuss, is certainly not the only one of its kind—thereby also addresses the question of why, in an age replete with fiber-optic cables and liquid-crystal displays, people are drawn to technologies and objects that refute trends toward perfection, transparency, and clarity.

It might be tempting to trivialize such sentiments as simple romanticisms in a digitalized world, or, more critically, to view them as symptoms of the stifling ubiquity of nostalgia within capitalist cultures (Reynolds 2011; Fisher 2014). One might, moreover, point out that it is not uncommon for people to cling to the mundane in times of technological upheaval, to glorify the outmoded as "an instinctive impulse to put the brakes on technological progress" (Reynolds 2011: 351). Yet, Peel's words complicate such readings. What is pertinent is that Peel presents a technological aesthetic of imperfection as desirable not so much for its connotations of obsolescence or for the familiar comfort it brings, but rather for how it speaks to the overarching conditions of existence; within vinyl's fragile properties, Peel feels the vibrant yet precarious pulse of life.

These observations delineate the scope of this book, which is guided by two primary concerns. First, I seek to unravel the links between imperfection as a *material, aesthetic category* and imperfection as an *existential condition*. Why, I ask, are aesthetic imperfections so frequently framed as emblematic of a wider logic of imperfection—a logic that thoroughly suffuses existence? Second, I question what the cultural significance of a technological aesthetic of imperfection might be in the contemporary era of what I define as *frictionlessness*, indicating a pervasive technological design philosophy that aesthetically glorifies connectivity, transparency, and user-friendliness while bringing a range of destructive effects. Concretely, I ask: If an aesthetic of imperfection can emotionally alert its beholder to the fragile conditions of existence, how might this (help to) challenge the more troubling ramifications of the current fetishization of technological perfectibility? To answer the first question, Chapter 1 presents a theoretical framework that links the aesthetic, existential, and ethical dimensions of imperfection. To answer the second question, the book traverses a sprawling terrain of technological imperfections, assessing digital glitches and lost transmissions, sonic malfunctions and obsolete devices, truncated connections and thwarted technical processes. By contrasting the aesthetic experiences to which technological imperfections may give rise with the aesthetic logic of what, in Chapter 2, is conceptualized as the design philosophy of frictionlessness, I develop an argument about how technological aesthetics of imperfection may serve as a counterweight to some of the most problematic cultural and environmental tendencies of today's digital technologies. While I reserve an in-depth engagement with the concept of imperfection for the next chapter, I will use this introduction to briefly sketch out some of the defining qualities of imperfection and to demarcate my usage of and interest in the term.

Imperfection

The word "imperfection" refers to the state of being imperfect. It is etymologically derived from the Latin word *imperfectus*, meaning "not complete," and, as the Cambridge Dictionary (2022) spells out, is generally used to label objects, conditions, or practices that are in some way faulty, fragile, or finite. The word is defined by an element of negativity: im-perfect means *not* perfect. It follows that imperfection is always related to a corresponding idea of perfection. Any serious inquiry into the notion of imperfection, then, must also grapple with the various conceptions of perfection against which the term is construed. Another curious quality of the label of imperfection is that, for all its seemingly depreciative implications, it regularly elicits an affirmatory response. As Peel's anecdote shows, the negative connotations of imperfection can be reconfigured as a positive testimony of the vagaries of existence. In its deviation from some established norm of perfection, an aesthetic of imperfection is frequently hailed as fertile, vibrant, and life-affirming (see, for example, Rutten 2021; Kelly, Kemper, and Rutten 2021). Moreover, while imperfection at face value implies a lack, it is in fact its antonym "perfection" that is often found wanting; for example, Dutch designer Hella Jongerius states that "'perfection'... renders objects emotion-proof" (quoted in Antonelli 2010: 233), signaling that, in its purported purity and fullness, perfection is regularly seen to lack warmth, vitality, and sometimes even humanity.[1]

My research into the notion of imperfection has materialized under the auspices of Sublime Imperfections—an NWO-funded, Amsterdam-based research project led by cultural historian Ellen Rutten and further made up of art theorist Fabienne Rachmadiev and myself. Sublime Imperfections departs from the observation that, in the past two decades, there has been a proliferation of interest in the imperfect, describing a preoccupation with objects and practices that in some way challenge norms of perfection and a tendency to frame such phenomena as valuable and appealing (Rutten, forthcoming). From the sun-bleached and sepia-stained granularity of prominent Instagram filters to the normalization of glitch-based effects in popular music and from the high demand for vintage clothing to the recent occurrence of the "vinyl revival" (Devine 2019: 9): indeed, many contemporary cultural practices exhibit a desire for the aesthetically imperfect. However, while the Sublime Imperfections project is concerned with these more pervasive and commercialized expressions of imperfection—often related to a commodification of nostalgia and authenticity—there is a more specific (and, admittedly, frequently more niche) form and quality of imperfection that I am interested in here. As illustrated by Peel's anecdote, my focus is

---

[1] The impulse to equate imperfection rather than perfection with the human condition has a long history, one that is interwoven both with religious worldviews—in which the human is often framed to comprise but a faint echo of the transcendent perfection of the divine (see, for example, Grant 2012)—and with cultural conceptions of technology—see, for example, political theorist Thomas Rid's identification of a pervasive "tendency to think about the machine in perfection, not in limitation" and, by implication, about the human as imperfect and error-prone (Rid 2017: 350–1). In the face of idols both religious and technological, imperfection has thus long been associated with the volatile business of being human. For further reading on the (theological and secular history of the) concept of perfection, see Hartshorne (1962); Foss (1946).

on aesthetics of imperfection that stimulate an emotional attachment to technology, particularly through the technological materialization of shared conditions of finitude and fragility.

I thus scrutinize but some of today's many popular expressions of imperfection. Within digital culture alone, there are numerous (at times conflicting) trends in which users and creators exhibit an appreciation of the aesthetically imperfect. There is, for instance, a growing tendency for content creators and influencers to opt for a raw and unfiltered style of editing, embracing an ethos of imperfection to appear more relatable (Reade 2021). Here, an imperfection-oriented style attains a certain symbolic value that appeals to (perceived) ideas of authenticity and self-exposure. However, as indicated, in what follows I will be less concerned with the widespread appearance of carefully curated expressions of imperfect and unfiltered content and more with imperfection as affectively encountered in technological objects and processes themselves. Specifically, I analyze contemporary technological objects that can be understood in terms of the imperfect as a result of their emotionally resonant negation of digital norms of permanence, perfectibility, transparency, and connectivity.

It is important to note that a (technological) affinity for imperfection is certainly not historically unprecedented (Rutten 2021). We should, moreover, be careful not to present such predilections as uniformly globalized phenomena (Saito 2017b). Nonetheless, it is productive to ask what alternate conceptions of technology an aesthetic of imperfection today enables. It is also critical to acknowledge that different expressions, discourses, and logics of imperfection may clash and override one another— the aesthetic, ethical allure of technological imperfection that I examine may certainly be affected or even diminished by different understandings and desires for the imperfect. However, unpacking the specific aesthetic power of technological imperfection is crucial for its culturally, socially, and environmentally significant tendency to directly undermine today's dominant, destructive design philosophy of frictionlessness. As I will argue, frictionlessness merges an aesthetic logic of technological perfectibility with a wealth of exploitative and extractive practices, necessitating the conception of aesthetic alternatives. In this regard, I hope my research will also serve as an invitation to theorists, artists, and designers to pay more attention to the ethical significance of aesthetic expressions of imperfection and to invoke imperfection's marked capacity to shape more care-inflected relations between humans and technology.

The project of Sublime Imperfections and my individual research are part of a growing field of imperfection studies (Rutten and de Vos 2023). Recent inquiries into the cultural potential of an aesthetic of imperfection—

or of related notions like noise, breakdown, and error—are found evaluating disciplines as varied as musical production (Brøvig-Hanssen and Danielsen 2016; Kelly 2009); digital design (Betancourt 2017; Cramer 2015; Kane 2019; Moradi et al. 2009); cinema (Rombes 2009); architecture (Hughes 2014); and photography (Chandler and Livingston 2012; Chéroux 2003). Like my own research, these studies explore the subversive and stimulating qualities that an aesthetic of imperfection holds, often in relation to today's digitalized mediascape. However, in contrast to the more discipline-specific outlook of most of these works, my study traces imperfections across different media, elucidating patterns and consistencies in the process.

While many of the aforementioned works present imperfection in an exclusively affirmative register, there have also been scholars who are more ambivalent about the ethical implications of appreciating the aesthetically imperfect. For example, while philosopher Yuriko Saito accentuates the value of an "imperfectionist aesthetics" that finds beauty in the decrepit, she is also sensitive to the more pernicious ramifications of such a perspective (2017b). Saito suggests that, when glorifying decay and dilapidation, one is at risk of glossing over the human misery such aesthetic attributes all too often signify: "There is something morally problematic about deriving an aesthetic pleasure from the sign of social ills suffered by others" (2017b). One salient example of the potentially harmful aspects of appreciating imperfection is found in the phenomenon of "ruin porn," or the cultural fetishization of urban sites of disrepair and pollution, which, as a pastime for the privileged, has been criticized for its "tendency towards exploitation and the trivialisation of economic struggle and decay" (Lyons 2018: 4; see also Radulescu 2021). As another ethically contestable example, Dutch designer Piet Hein Eek's famous scrap wood furniture calls to mind a long legacy of poverty-stricken beachcombers who had to carve out an existence from whatever the sea would freight landwards. Eek's furniture, in contrast, sells for thousands of dollars. These examples, however, do not mean that an appreciation of imperfection, even if it coalesces around run-down objects, should automatically be condemned. As Saito argues, completely rejecting an imperfection-oriented aesthetic for its problematic implications would close off an entire realm of possible aesthetic experience and would thereby also curtail the many potential benefits such an aesthetic might bring (2017b). Rather, aesthetic appreciations of imperfection should be attended by a careful consideration of their object and the social and cultural conditions that have informed it.

In my interest in unraveling the significance and potential of a technological aesthetic of imperfection, I follow Saito in affirming the "considerable power of the aesthetic to guide people's behavior, decisions, and actions" (2017a:

141). The political and ethical stakes of this transformative power are further emphasized by environmental scholars Robert S. Emmett and David E. Nye: knowledge, they surmise, must be "*affective*, or emotionally potent, in order to be *effective*, or capable of mobilizing social adaptation," and aesthetics provides one way of rendering objects and practices emotionally compelling (2017: 8, emphasis in original). Both of these claims stress the possible value of an aesthetic of imperfection, even if it represents negative or hazardous phenomena. Moreover, they indicate that the ability of aesthetic imperfections, already hinted at by Peel, to point to the fragile conditions of existence can also serve to spark an investment in the way these conditions unfold materially. It might, for instance, very well be that a record like ambient producer Rafael Anton Irisarri's *Solastalgia* (2019), whose melancholic aesthetic of imperfection enacts the vanishing of familiar worlds, much more pressingly communicates the perils of climate change than does the frigid objectivity of the scientific report. Similarly, as artist/researcher Susan Schuppli's discussion of filmmaker Vladimir Shevchenko's corrupted footage of Chernobyl conveys, flawed and eroded imagery of historical disasters can elicit emotions that prove more responsive to the depicted tragedy (2020: 61–5). Emphasizing rather than resisting the aesthetic imperfections such damaged objects betray can attune viewers to the troubling circumstances of their genesis and can become "ethically productive and evidential of a moral truth" (Schuppli 2020: 38). These ethically charged examples show that an aesthetic of imperfection, in its materialization of conditions of finitude and fragility, can encourage a more careful and compassionate attitude in its audience.

Research Setup and Outline of the Book

As stated, I will primarily be concerned with the value of these critical and ethical qualities in the context of contemporary technology. How, I ask, does a technological aesthetic of imperfection (or an aesthetic of imperfection as it is found in technological objects and processes) serve as a counterweight to the perfection-oriented logic of frictionlessness that reigns today? The design philosophy of frictionlessness, as will be further explicated in Chapter 2, poses an aesthetic demand for spotless technological designs that function ever more transparently. While it might appear auspicious, frictionlessness is far from a benign philosophy, as its technologies require and exploit vast networks of vulnerable and finite resources—a spectral underside that is aesthetically hidden from the user's gaze. Saito maintains that we should "critically [analyze] the aesthetic effects of familiar objects and environments in our lives" and should question how to create "positive aesthetic effects through designing objects and environments and practicing respectful and caring interactions" (2017a: 225). I adopt such a critical perspective in the context of the digital, interrogating, on the one hand, the aesthetic effects that emerge from the pervasive philosophy of frictionlessness and, on the other hand, the possible challenges a technological aesthetic of imperfection may stage.

What my study primarily aims to add to the existing body of work on imperfection and digital culture is, first, a sustained theorization of the links between imperfection as an aesthetic category and imperfection as an existential condition. As some of the preceding examples show, there is, for many people, a frictional quality to an aesthetic of imperfection that speaks to the shared conditions of existence and that throws the contingencies that come with navigating life into sharp relief. However, in untangling the links between imperfection in an aesthetic and an existential register, this study aims not only to chart the appeal of the aesthetically imperfect but also to consider the critical or subversive power of imperfection in the face of today's technological culture of perfectibility. Regarding this last dimension, this study offers the first sustained conceptualization of frictionlessness as the dominant digital design philosophy of our time. While the terms "friction" and "frictionless(ness)" are regularly mobilized by scholars trying to make sense of the tech sector's designs—media theorist Jathan Sadowski, for instance, designates "frictionless" to be the buzzword of Silicon Valley (2020: 47)—this has not yet resulted in a comprehensive account of frictionlessness as the driving force behind (consumer) technology. I conceive of frictionlessness as a perfection-oriented design philosophy whose aesthetic logic carries a range of toxic effects. Ultimately, I develop an argument about the value of a technological aesthetic of imperfection in addressing the existential and environmental ramifications of frictionlessness as a digital mode of production and consumption that operates precisely by concealing its own destructive nature.

Methodologically, I take cues from the fields of cultural analysis and postphenomenology. From cultural analysis, I borrow a focus on interdisciplinarity (combining a media studies perspective with insights from cultural theory, continental philosophy, media archaeology, and the environmental humanities) and on the synthesis of theoretical development and meticulous analysis of cultural expressions (Bal 1996). Moreover, I adhere to the focus of cultural analysis on the contemporary moment—rather than seeking to historicize the notion of imperfection, I am interested in unraveling how technological imperfection matters *now*, in a time paradoxically defined by advanced technological sophistication and growing inequality and environmental catastrophe. I combine the outlook of cultural analysis with a postphenomenological approach. Postphenomenology, whose theoretical and methodological qualities I flesh out in Chapter 1, is premised on the notion that technology forms an indelible part of how humans relate to the world. It is concerned with "conceptual analysis of the implications of technologies" and takes "empirical work as a basis for philosophical reflection" (Rosenberger and Verbeek 2015: 31). In conjunction, the approaches of cultural analysis and postphenomenology—combining close attention to cultural expressions with the assertion that technology is central to, but not unilaterally determinative of, the phenomena under investigation—provide the methodological scaffolding that allows me to answer my guiding questions.

In terms of organization, Chapter 1 construes a theoretical foundation for the rest of the book, Chapter 2 advances a conceptualization of the philosophy of frictionlessness and Chapters 3 through 5 follow the empirical emphasis of cultural analysis and postphenomenology by presenting close readings of several technological objects that exhibit a distinct aesthetic of imperfection. In Chapter 3, I analyze the Dutch video game *GlitchHiker*, in which an aesthetic of imperfection was used to communicate the game's imminent demise, spurring players to become emotionally invested in the finite existence of this digital object. In Chapter 4, I discuss Rosa Menkman's *The Collapse of PAL*, an audiovisual performance that mobilized an array of glitches to stage a critique of cultural norms of technological obsolescence. In Chapter 5, I conduct a reading of the musical genre of vaporwave and particularly the work of Dutch vaporwave producer Cat System Corp. The sound of technological deceleration and imperfection that defines Cat System Corp.'s work, so I suggest, serves to dramatize the experiences shaped by today's technologically ordained consumerism. Each of these objects teaches us something about the significance of technological imperfection in the face of a widespread culture of digital perfectibility.

In investigating objects made and mostly viewed in a Western context, I adopt an admittedly limited view. This view is largely restricted to that of the Western user/consumer whose use of technology relies on a global chain of destruction and exploitation. However, following earlier arguments that aesthetics play a key role in facilitating emotional attachments, part of my argument about the value of a technological aesthetic of imperfection attempts precisely to widen the narrow scope of the user's/consumer's gaze: countering current aesthetic regimes, a technological aesthetic of imperfection can facilitate different ways of relating to technology—ways that perhaps prove more sensitive to the many lives and matters that are at stake in today's technological practices of production, consumption, and disposal.

It is also worth noting that the objects I analyze have all been made by Dutch artists/creators, but that this marks a level of local specificity I engage only modestly. To be sure, there is plenty of scholarly work that sees a distinct link between imperfection and Dutch art and design: researchers have, for example, traced the Dutch artistic predilection for aesthetic imperfections back to the "Calvinist national character" that traditionally relishes unpolished styles (Junte 2011: 7); the Dutch frugal spirit, which inspires the recycling of old products and materials (Thomas 2008: 226); and the expansion of a high-tech industry that has nonetheless retained a pronounced focus on handicraft (Junte 2011: 15).[2] These are dimensions that other scholars of my analyzed objects may well choose to underscore, but in reading these objects I was not primarily concerned with their "Dutchness." For one, this is because my theorization of frictionlessness as a pervasive philosophy is focused more on its generic aspects and less on national particularities, but also because the aesthetic of imperfection that marks my objects reveals or engages technological processes that extend beyond national boundaries. The empathic responses that *GlitchHiker* elicited were, in my view, less a result of any specifically Dutch property and more of an appeal to the universal human capacity to be affected by loss. *The Collapse of PAL* criticizes a logic of technological optimization and obsolescence that transcends national borders; the performance, commissioned for Danish rather than Dutch television, decries the discontinuation of a particular technology in a host of different countries rather than in one restricted locale. The genre of vaporwave to which Cat System Corp.'s work belongs presents a global narrative of ubiquitous consumerism; the genre probes the similarities rather than the differences between consumerist societies and, while I will be critical of interpretations that gloss it as truly globalized, endeavors to evoke the transnational fluidity of capital. Each of these objects, in sum, emphasizes the Western ubiquity and confluence of technological processes of obsolescence and consumption rather than the specificities of Dutch (digital) culture.

Chapter Outline

In Chapter 1, I develop a theoretical framework that enables me to conceptualize imperfection as both an existential condition and an aesthetic quality. I do so by engaging the work of three philosophers: Jacques Derrida, Martin Hägglund, and Bernard Stiegler. The first part of

---

[2] For further reading, see Thomas (2008); Vlassenrood (2009).

the chapter introduces four concepts that are central to the entire study: *autoimmunity, the existential primacy of imperfection, chronolibido*, and *spectrality*. Autoimmunity, a concept developed by Derrida and further explicated by Hägglund, marks the co-implication of time and space and discloses that all things must bear within themselves the germs of their own dissolution. This constitutive condition also reveals what I term the existential primacy of imperfection, or the (onto)logical impossibility of absolute perfection. I then move on to discuss Hägglund's concept of chronolibido. Chronolibido charts the co-implication of finitude and desire, revealing that desire is always linked to the possibility of loss, that the existential primacy of imperfection therefore forms the precondition for all forms of care, and that investments of care can be elicited through an aesthetic of imperfection. As another central concept, I introduce the Derrida-inspired notion of spectrality, which serves a twofold function. First, it demarcates how a logic of haunting pervades existence, further specifying the implications of the condition of imperfection. Second, it registers the materially specific ghostly effects that this spectral logic makes possible, allowing one to assess how past and future, absence and presence, and life and death are materially active in a historically, technologically situated present. This latter dimension informs my reading of frictionlessness in Chapter 2.

The second part of Chapter 1 introduces the work of Bernard Stiegler, whose account of the technological foundation of human time-consciousness, described by him in terms of *tertiary retention*, captures the inextricable convergence of technology and human consciousness. Stiegler shows how, in addition to chronolibidinal beings, we are also *pharmacological* beings: beings whose temporal consciousness is by default malleable by technology, which, as a result, can always produce both curative and poisonous effects. I argue that one crucial pharmacological dimension that Stiegler does not sufficiently explicate is that an individual's ostensibly beneficial relation to technology can have a pernicious impact on the existence of others. This dimension should urge us, I propose, to also consider how technological aesthetics alert the user to all the other lives and matters that are at stake in technology's production. As my later case studies will show, imperfection, as an aesthetic category that emotionally binds users to existential conditions of finitude and fragility, offers one possible avenue for finding more sustainable, chronolibidinal relations to technology itself. I conclude the chapter by underlining why it is important that humans retain responsibility over the mobilization of technology; the ability to recognize finitude and imperfection is what distinguishes us from our machines and should thus be emphasized within human–technology interactions.

Chapter 2 presents a pharmacological reading of the pervasive technological design philosophy of *frictionlessness*, focusing primarily on the toxic ramifications of its aesthetic logic. Frictionlessness is first conceptualized as an aesthetic collusion of surveillance capitalism and platform capitalism. Then, it is argued that this philosophy shapes a user experience that combines the values of user-friendliness, connectivity, and optimization; frictionlessness maintains that the perfectibility of consumer technology lies in designing appliances that function so smoothly and that are woven so seamlessly into the fabric of everyday life that technology progressively recedes from perception while mediating an increasing sphere of human activity. I propose that the aesthetic logic behind this philosophy can best be characterized by identifying three specific forms of spectrality that guide it: an increase in *technological spectrality*, an intensification of processes of *spectralization*, and the delimitation of *hauntological aesthetics*. Through increasingly hiding its operations from the user, frictionlessness further exploits but also obscures the finite conditions of its production and discourages the aesthetic conception of alternate futures. As a possible counterweight to the ghostly logic of frictionlessness, I conceive of a technological aesthetic of imperfection as a source of *friction* that potentially activates its audience into more sustainable modes of engagement, illuminated by a careful sensitivity to conditions of finitude and fragility. The chapter concludes that, in the destructive era of frictionlessness, the human capacity to intervene on the basis of a recognition of the existential primacy of imperfection should be pharmacologically accentuated rather than further negated. Imperfection is conceptualized as one aesthetic category through which perception can be reconstrued in a more pharmacologically sustainable fashion.

Chapter 3 comprises the first of my three case studies: Vlambeer's video game *GlitchHiker* (2011). As the chapter elaborates, *GlitchHiker* was a game that was programmed to die and a playable version of the game thus no longer exists. *GlitchHiker* communicated its mortal state to the player through a *glitch*-based aesthetic of imperfection that became more pronounced as its end approached. I explain that glitch is a concept that describes notable flaws in technological operations and that it is often labeled as both an imperfection and a ghost in the machine. The chapter demonstrates that this ghostly essence derives from three central qualities: glitch's capacity to highlight the unseen, glitch's capacity to indicate an unassailable technological agency, and glitch's capacity to spell technological death. It is primarily this last capacity that helped to make *GlitchHiker* such a unique event; as its creators quickly found, players began to care for the moribund game in a way that diverges from routine modes of engaging technology. I analyze this phenomenon through the lens of Martin Hägglund's notion of chronolibido and Steven Jackson's studies of repair and maintenance. *GlitchHiker*, I demonstrate, reveals the pertinence of chronolibido to technology, bolstering visions of a technological culture that restores rather than replaces and that mends rather than rejects.

Chapter 4 focuses on glitch artist/theorist Rosa Menkman's audiovisual performance *The Collapse of PAL* (2010). This performance laments the fate of a terminated technology, the PAL (Phase Alternating Line) signal, that was replaced by a digital, more frictionless alternative. Through an aesthetic of imperfection and a narrative that summons Walter Benjamin's canonical Angel of History (a figure that Benjamin invoked to condemn the destructive underside of progress), *The Collapse of PAL* urges its audience to care for its dead, even if these dead are technological in nature. Through a close reading of Menkman's performance, I develop three central arguments: (1) the figure of the ghost, when measured against the oft-used concept of zombie media, offers a more productive lens through which to understand the afterlives of technologies, with their complex diffusion of environmental effects; (2) *The Collapse of PAL*, in its reimagination of Benjamin's Angel of History, exemplifies a historico-aesthetic practice of seeking in technological aesthetics of imperfection traces of the destructivity of frictionlessness; and (3) through exhuming the disused, Menkman helps us to think through new modalities of caring for the realm of technology. Such modalities are necessary, I contend, if we are to cultivate more sustainable forms of technological interdependency.

Chapter 5 probes the musical genre of *vaporwave*, and particularly the work of Dutch vaporwave producer Cat System Corp. I first analyze vaporwave as an audio(visual) genre of electronic music that is aesthetically premised on imperfection, spectrality, and the signs and sounds of consumer capitalism. This genre, characterized by a heavy use of loops, reverb, and the deceleration of pop songs, enacts a rhythm of what Simon Reynolds calls *hyper-stasis*: a paradoxical amalgamation of technological acceleration and cultural stasis. I further develop this argument through a close reading of Cat System Corp's work, particularly his album *Palm Mall Mars* (2018b). Cat System Corp's work, I demonstrate, speaks to Bernard Stiegler's notion of *tertiary retention* and reveals how, within consumerist societies, the human mind has become emotionally tied up with technological signs of consumption. Vaporwave, more exactly, dramatizes a mind that is thoroughly haunted by its own consumerist past. By relating vaporwave to two of the forms of spectrality I discussed in Chapter 2, I argue that this musical genre is indicative of the dominant relation to consumption that defines societies marked by the designs of frictionlessness. I conclude the chapter by meditating on the profound quality of vaporwave to reappropriate technological objects and to stimulate different engagements with technology, but also signal that there are limits to the genre's critical potential.

In the Coda, I introduce the notion of technological melancholia as a temperament that forms an appropriate response to the pharmacological effects of frictionlessness. Melancholia describes a sentiment that refuses to lay its specters to rest and thus appears pertinent in a technological time that produces so many marginalized ghosts. The Coda offers a reflection on the capacity of a technological aesthetic of imperfection to elicit feelings of melancholia, and on the implications of such a sensibility in relation to the poisonous drives of frictionlessness. I conclude that the path to a more technologically sustainable future can be found only by making room for the vulnerable, the liminal, and the dead. The Coda brings together the findings of my framework and case studies, accentuating the profound capacity of an aesthetic of imperfection to spark care and concern for the existentially universal—but always also materially and technologically specific—unfolding of finitude and fragility.

1

The Existential Primacy of Imperfection

Autoimmunity, Chronolibido, Spectrality, and Technology

"Imperfection" is an elusive and polysemic term that manifests itself in a myriad of disciplines and that is mobilized in a number of different ways. In the Introduction, I suggested that this heterogeneous term is nonetheless defined by a tenacious motif: imperfection as an aesthetic trait is frequently seen as symptomatic of imperfection as a general condition of existence. To make sense of this double register in which imperfection operates, this chapter develops a theoretical framework that engages imperfection both as a primordial logic that defines all that exists and as an aesthetic quality. I understand imperfection as existentially primal insofar as it is descriptive of the *a priori* conditions of existence. Subsequently, I approach imperfection in its aesthetic manifestations, and especially in the context of technology, as a means of unveiling and attuning people to (the material effects of) these overarching conditions. The arguments developed in this chapter are primarily built on the work of three philosophers: Jacques Derrida, Martin Hägglund, and Bernard Stiegler. While Hägglund and Stiegler are both Derridean philosophers who have inherited Derrida's spirit of deconstruction, they do not draw on one another and have not yet been read in relation to each other. This chapter, in its analysis of imperfection, synthesizes their work and shows how they complicate, but also complement, each other's thinking, especially in relation to matters of technology, finitude, and phenomenology.

I begin this chapter by introducing the notion of *autoimmunity*, not as a biological phenomenon but as a philosophical concept advanced by Derrida and further developed by Hägglund, to theorize imperfection as a constitutive logic that pervades existence, preceding all material manifestations of imperfection. This theorization discloses what I call the *existential primacy of imperfection*, or the (onto)logical impossibility of perfection. I subsequently discuss Hägglund's notion of *chronolibido*. Chronolibido denotes a theory of care and desire that foregrounds the human perception of the existential logic of imperfection, allowing me to approach imperfection as both an aesthetic quality and an existential condition. Consequently, I return to Derrida to address the concept of *spectrality*, another constitutive condition that helps to explain how the workings of imperfection play out materially. By briefly reflecting on the field of postphenomenology, I underline how technology necessarily mediates our relation to the preceding concepts. The identification of this inevitable mediation leads to a critique of Stiegler's conceptualization of the *pharmakon* (technology as always both poison and cure). Reorienting Stiegler's interpretation of the *pharmakon* is crucial, I contend, for understanding the value of an aesthetic of imperfection today, within the here and now of digitally transformed societies. I close the chapter with a discussion of the significance, within the fold of human-technology interaction, of the human awareness of the existential primacy of imperfection: this capacity is what should urge us to remain responsible for the way technologies develop. Technological aesthetics can, I conclude, play a crucial role in cultivating such an attitude.

Autoimmunity and the Existential Primacy of Imperfection

I opened the Introduction with a brief account of John Peel's preference for the noisy imperfections of vinyl for their approximation of the noisy venture that life itself is. Peel's words, I suggested, are indicative of a tendency to frame materially, aesthetically specific instantiations of imperfection as emblematic of imperfection as a universal condition of existence. Jacques Derrida's concept of *autoimmunity* offers valuable resources for grasping how existence is necessarily haunted by such a sense of imperfection. "Autoimmunity" is a term that began to appear only in Derrida's later work—one encounters it, for instance, in his conversation with philosopher Giovanna Borradori, where he defines it as "that strange behavior, where a living being, in quasi-*suicidal* fashion, 'itself' works to destroy its own protection" (quoted in Borradori 2003: 94, emphasis in original). The most sustained exposition of the concept comes, however, not from Derrida himself, but from philosopher Martin Hägglund's reading of him. While, as stated, the term "autoimmunity" materialized only at a late stage of Derrida's output, Hägglund demonstrates that the logic it bespeaks has, in fact, always been central to Derrida's thinking. Hence, in developing the argument that autoimmunity affords a theoretical entryway toward a better understanding of imperfection, I will draw not only on Derrida's own words but also on Hägglund's interpretation of them. Reading the concept of autoimmunity in relation to imperfection will allow me to make sense of how imperfection and perfection are logically opposed and will furthermore disclose how an ineluctable logic of finitude ensures imperfection to be a foundational condition.

Hägglund gives the following succinct description of the Derridean notion of autoimmunity: "Autoimmunity spells out that everything is threatened from within itself, since the possibility of living is inseparable from the possibility of dying" (2008: 9). The far-reaching implication of this claim is that, for Derrida, there is absolutely nothing that can be wholly indefectible in its being. This places Derrida at odds both with philosophical traditions that hypothesize a transcendence beyond time and space and with religious conceptions of God as a figure of the infinite. Derrida's seminal 1967 text on Emmanuel Levinas, "Violence and Metaphysics," is particularly illuminating here, because it teems with traces of the logic of autoimmunity. In this text, Derrida describes how

> the thought of Being, in its unveiling, is never foreign to a certain violence. That this thought always appears in difference, and that the same— thought (and) (of) Being—is never the identical, means first that Being is history, that Being dissimulates itself in its occurrence, and originally does violence to itself in order to be stated and in order to appear. A Being without violence would be a Being which would occur outside the existent: nothing; nonhistory; non-occurrence; nonphenomenality. (1978b: 184, emphasis in original)

As Derrida implies, "violence"—a term I will soon further explicate—is not inimical to but constitutive of existence.

The concept of autoimmunity registers this constitutive condition and conveys, simply put, that all things bear within themselves the seed of their own dissolution. A prime instantiation of this logic is life: the very code of life has death written into it and all living beings thus edge toward their inevitable demise. Yet, the logic of autoimmunity exceeds the notion of (biological) life as such; it is, following Derrida, foundational to *all* that exists, affecting organic and inanimate matter alike. Autoimmunity also defines, for example, the realm of technology; no technology exists that could, even in theory, remain unmarked by finitude and disorder (Derrida 2005: 22–3; see also Bennington and Derrida 1993). Processes of memorization, transcription, and archivization are equally formed by an incontrovertible fragility; these practices would not make sense if the possibility of their undoing were not already latent. The archive, so Derrida maintains, "always works, and *a priori*, against itself" (1995: 12, emphasis in original) and is necessarily conditioned by "the violence of forgetting" (1995: 79; see also Derrida and Stiegler 2002:

51). Autoimmunity, in sum, reveals how a logic of finitude is not something imposed from the outside. It is, rather, always-already "requisite in the very structure that it solicits" (Derrida 1982b: 133).

Yet, what decrees this inescapable finitude, clawing away from the inside? As Hägglund incisively illustrates, it is the spacing of time (or *espacement*), as conceptualized by Derrida, that accounts for autoimmunity. The logic of spacing is also what determines the nature of imperfection, so it is important to briefly unpack it here. Channeling Derrida, Hägglund states that "any so-called presence is divided in its very event and not only in relation to what precedes or succeeds it" (2008: 17). This means that each moment passes away as soon as it comes into being and that time, therefore, continually negates itself. The only way that a moment can thereby remain for the future is by inscribing itself *spatially*, as a so-called *trace*. The trace, which as a concept is pivotal to Derrida's entire project of *différance* and deconstruction,[1] represents the logical co-constitution of time and space. As such, it describes the *a priori* condition of all material traces, whether this concerns the most ancient of fossils or the most fleeting of melodies. Traces link past, present, and future through the movement of spacing, which Derrida describes as the becoming-space of time and the becoming-time of space. The "now" must in some way render itself spatially in order not to succumb to time's self-negation, as spatiality is what can remain over time (hence, the becoming-space of time). Through this gesture of spatialization, time is made dependent on a material carrier and, therefore, each "element appearing on the scene of presence, is related to something other than itself" (Derrida 1982a: 13). Spatiality, however, is also always-already marked by temporality—an inscribed trace cannot be apprehended immediately at its inception precisely because space is traversed by time, and traces are thus always left for a future that is ever uncertain (hence, the becoming-time of space). This co-implication, necessarily engendering a movement of both spatialization and temporalization, or spacing for short, is what Derrida posits to describe the "'originary constitution' of time and space" (1982a: 8).

What is crucial about the concept of autoimmunity is that this originary constitution, forming "the condition for everything *all the way up to* and including the ideal itself . . . [and] . . . for everything *all the way down* to the

---

[1] Both of these canonical Derridean concepts describe the operations of spacing. *Différance*, as the relentless movement of difference and deferral, derives from spacing's necessary logic of alterity. Deconstruction, as a way of exposing and overturning the "violent hierarchy" that structures all texts (Derrida 1981b: 41), is another notion that is premised on spacing's fundamental negation of stability.

minimal forms of life" (Hägglund 2008: 19, emphasis in original), opens all that exists up to the risk of violence. Violence should here be understood not as a necessarily harmful or malign infringement, but rather as an openness to finitude and alterity that is necessitated by the impossibility of self-containment. Autoimmunity defines an internal logic of fragility that ensures that nothing can be fully inoculated against being affected by what is other than itself. The minimal condition for a trace to exist is that it joins in the movement of spacing and this movement is always underpinned by the violent possibility of change. A trace, in this regard, is exposed to forces that can never be decisively controlled (the very sentence I just wrote can be erased, rewritten, misunderstood). Spacing entails that every generative gesture of inscription must already be haunted by death, as to be granted the possibility of survival is also to be consigned to the unwavering law of finitude: "Traces thus produce the space of their inscription only by acceding to the period of their erasure" (Derrida 1978a: 284). It is this postulation of a constitutive openness to mutability, of an ineradicable possibility of violence and contamination, that informs the entirety of Derrida's deconstructive endeavors. Derrida's theorization of the ineluctable spacing of time thereby challenges the coherence of any worldview sustained by an active or suppressed belief in the prospect of purity, presence, transcendence, or infinitude.

What does this mean for my consideration of imperfection? Above all, it shows that absolute perfection is, (onto)logically speaking, impossible. A claim of perfection would effectively amount to a declaration of the abolishment of time. Time, due to the movement of spacing, always allows the violent possibility of alteration and that which is perfect would not permit this contaminative risk; otherwise, it would not be perfect. By this measure, unconditional perfection is a logical impossibility. Hägglund confirms this claim when he explains how the logic of autoimmunity "undercuts the regulative Idea of final perfection. The impossibility of such an absolute state is not a privation but the possibility of change at any juncture, for better and for worse" (2008: 169). Even if a state of perfection is seemingly reached, this state remains temporal and can therefore always be altered.

The work of Derrida and Hägglund thereby discloses what I will refer to as the *existential primacy of imperfection*. This term describes how autoimmunity and the spacing of time ensure the ontological impossibility of perfection, understood here as a structural integrity that would be fully unscathed and immutable. To conceive of imperfection as existentially primal is not a means of positing a primordial logic absolved from time and space, but a way of specifying the materially constitutive conditions that result from the necessary synthesis of temporality and spatiality. The universal condition of imperfection engendered by this synthesis signifies the impossibility of transcending the temporal (because temporality is not a cast that could be shed, but the source of all existence), of fully immunizing oneself against change (because autoimmunity spells an internal foundation of finitude and fragility), and of attaining ultimate control (because the spacing of time always ordains the coming of an uncertain future). These impossibilities cannot be revoked by material circumstances and invalidate the chance of a culminating perfection.

One might rightfully point out that the preceding descriptions touch on some qualities that, in the Introduction, were ascribed to an *aesthetic* of imperfection, such as the representation of fragility and impermanence. Yet, it is important to stress that autoimmunity and the existential primacy of imperfection describe conditions that precede material phenomena and that no aesthetic injunction organically flows from them. My argument is thus not that this existential primacy would in itself somehow stipulate a desire for the aesthetically imperfect. An aesthetic of imperfection can, however, make palpable or even dramatize the condition of autoimmunity and its negation of perfection. To phrase this more concretely, an aesthetic of imperfection can heighten one's sense that there is no escaping finitude, that fragility therefore describes a collective and ineradicable condition, and that perfect control can never be achieved. A prime example of the dramatic enactment of autoimmunity through an aesthetic of imperfection is found in artist/researcher Rosa Menkman's *The Collapse of PAL* (2010), an audiovisual performance that I analyze in Chapter 4. *The Collapse of PAL* is based on the contention that the technological quest for perfection is "a regrettable, ill-fated dogma" because all technologies have their "inherent fingerprints of imperfection" (Menkman 2011a: 11). Menkman here effectively describes the workings of autoimmunity: because of the primacy of imperfection, there is, indeed, no technology that could function infinitely or entirely without flaws. *The Collapse of PAL* amplifies this situation by confronting its audience with an array of technological glitches and failures, drawing on an aesthetic of imperfection to stage a narrative of technological finitude and obsolescence. The performance, then, leverages technological aesthetics to convey the imperfect condition of all technology. The aesthetic and emotional appeal of *The Collapse of PAL* also raises the more general question of how humans perceive and relate to the existential conditions set forth so far, and of the role technology has to play here. In addressing these questions, Hägglund's work again proves instructive: Hägglund's thinking on the relation between finitude and desire offers a starting point for analyzing the aesthetic significance of imperfection.

Chronolibido

At face value, the constitutive finitude that autoimmunity registers might seem like a terrible burden to bear. Is there a more harrowing tableau than the death of what we hold dear? Likewise, the idea that the memories that currently saturate our mind might one day be reduced to a pallid flicker, eventually to be quenched entirely, presents an utterly painful prospect. However, one of Hägglund's most remarkable contributions to Derridean philosophy is that he demonstrates how the autoimmunity-ordained possibility of loss— lamentable though it may appear—forms the condition for any emotional investment to take shape in the first place. This is made manifest by his concept of *chronolibido*. What initially drew me to this concept was my inquiry into *Glitchhiker*, a video game that was designed to expire and that I discuss as one of my three case studies. As will be elaborated in Chapter 3, *Glitchhiker* fostered an attitude of care in its players precisely because it was at risk of being irretrievably lost and communicated its finite state to the player through an aesthetic of imperfection. Chronolibido helps to account for this phenomenon; as a concept, it is key to understanding human relations to both the existential condition of imperfection and aesthetic invocations of imperfection that bespeak the reality of loss, fragility, and finitude.

As a prelude to Hägglund's concept, it is enlightening to consider the classic phrase of "*Et in Arcadia ego*," attributed to the ancient Roman poet Virgil. The translation of this phrase is "even in Arcadia am I," where the "I" is often taken to refer to the specter of death (Bowring 2015: 96). Arcadia was another name for paradise and the saying thus served as a reminder to "those in search of a *perfect* life, an ideal existence" (Bowring 2015: 95, emphasis added) that nothing can be exempt from finitude; even in paradise, death lurks. If one follows this interpretation of Virgil's dictum, one can read it as a representation of the condition of autoimmunity. It makes plain that even the most seemingly ideal state is always-already threatened from the inside; death is not displayed as encroaching on Arcadia from outside, but as already contained within. A chronolibidinal reading of the phrase would go even further, however, as it would maintain that death is a *necessary* part of whatever Arcadia we long for—autoimmunity and imperfection are not regrettable conditions that would ideally be overcome but are, instead, integral to the very structure of desire. The following passage from Derrida's work—parts of which Hägglund also cites (2008: 111)—underlines this:

> Ruin is not a negative thing. First, it is obviously not a thing. . . . One cannot love a monument, a work of architecture, an institution as such except in an experience itself precarious in its fragility: it has not always

been there, it will not always be there, it is finite. And for this very reason one loves it as mortal, through its birth and its death, through one's birth and death, through the ghost or the silhouette of its ruin, one's own ruin—which it already is, therefore, or already prefigures. How can one love otherwise than in this finitude? (2002: 278)

Here, Derrida is not advocating an ethical or aesthetic orientation that endorses dilapidation over immaculacy. Rather, he posits that the finitude that characterizes both the object of our love and the feeling of love itself is an indispensable precondition for the formation of any form of attachment. In everything we relate to, there is a necessary haunting of the present by the future; for any sort of affective bond to take shape, the ghost of impending ruin must, like a palimpsest of the future, already be legible in whatever we presently regard.

Hägglund elaborates these latter claims through the aforementioned concept of chronolibido. With this concept, Hägglund rereads the Freudian notion of libido and argues that care and desire are irrevocably embedded in finitude (2012: 3–4). He deconstructs philosophical and psychoanalytical notions, such as Freud's originary state of equilibrium and Lacan's "the Thing," that discern within desire a constitutive drive to overcome an ontological lack of some primordial or perfect state of being. By contrast, Hägglund maintains that desire is necessarily preceded and animated by the movement of spacing and by a foundational attachment to finitude. The concept of chronolibido that he develops from this premise is characterized by a double bind that incorporates the elements of *chronophilia* and *chronophobia*:

Chronolibidinal reading seeks to show that the ambivalence of desire stems from the double bind of temporal finitude. Desire is *chronophobic* since whatever we are bound to or aspire for can be lost: it can be taken away from or be rejected by us. Yet, by the same token, desire is *chronophilic*, since it is because we are bound to or aspire for something that can be lost that we care about it, that we care about what happens. (Hägglund 2012: 14, emphasis in original)

Only that which can be lost can be desired and, as such, *-philia* and *-phobia* co-imply one another. The fact that a moment passes forms the minimal condition for any desire to try to hold on to it. Taking away the possibility of loss would also render unintelligible the impetus to keep: "The fear of time and death does not stem from a metaphysical desire to transcend temporal life. On the contrary, it is generated by the investment in a life that can be lost. It is because one is attached to a temporal being (chronophilia) that one fears losing it (chronophobia)" (Hägglund 2012: 9). This also means that desire is informed by a preconditional investment in survival—survival as a finite being in time, affected by autoimmunity and, by implication, the existential condition of imperfection. Here, survival and finitude do not principally concern the extension or demise of biological existence but rather connote a constitutive attachment to temporality; even if one were immune to death and corporeal disintegration, one could still desire to retain a particular object, thought, or emotion, and would thereby remain entangled in the play of chronolibido (Hägglund 2012: 7–8).

This is not to say that people cannot in practice, to paraphrase the late musician/poet David Berman, "want the end of all wanting,"[2] but rather that the attachment to temporal finitude is integral to the structure of desire and precedes desire's concrete manifestations. Because the investment in finitude takes shape primordially to, and irrespective of, actualized desires, chronolibido is not a deterministic principle that dictates the content of our actions. Rather, it describes the condition of possibility for *all* forms of care and desire; we can want nothing, or can even yearn for death, but these nonetheless remain effects of the constitutive bond to the temporal (Hägglund 2012: 12–13).

The pure realm of perfection to which some metaphysical and religious worldviews maintain we aspire was already suggested to be unattainable (because of autoimmunity and the attendant primacy of imperfection), but Hägglund also reveals how such a realm would be stripped of all desire (because of desire's necessary link to temporality and loss). This imparts the unique relation to the existential primacy of imperfection that humankind has: our temporal sense of the ontological impossibility of perfection forms the basis of our very capacity to care and to desire. Our constitutive sense of being mired in the temporal forms the "fundamental trauma of chronolibidinal being" (Hägglund 2012: 152)—this trauma, however, does not derive from a primordial yearning for purity or perfection that furtively animates all desire, but rather emerges from the fact that "*pain and loss are part of what we desire*, pain and loss being integral to what makes anything desirable in the first place" (Hägglund 2012: 152, emphasis in original). This does not mean that pain and loss are in themselves desirable, or that we should not aim to reduce suffering, but that the spacing of time inextricably links the possibility of beauty to the promise of its destruction. One cannot eliminate the one without eliminating the other. As chronolibidinal beings endowed with an intuition of the temporal logic of autoimmunity and imperfection,

---

[2] This is a reference to Berman's 2019 song "That's Just the Way That I Feel," performed under the name of Purple Mountains.

we are therefore always doubly bound to finitude: the deepest love may turn to the most devastating grief, and this is what makes it love in the first place.

Hägglund expands his arguments through an exegesis of the literary works of Marcel Proust, Virginia Woolf, and Vladimir Nabokov by showing how, despite chronolibido's inherent lack of direction, it is possible to wrest specific aesthetic investments from the necessary conditions he specifies. He exposes how these authors, contrary to the purported desire to transcend time that is often ascribed to them, emphatically impart a wish to live on *in time*, fully admitting the possibility of violence this comprises.[3] Each of these authors distills a form of pathos precisely from aesthetically staging love and loss as inseparable. Hägglund demonstrates, for example, how Virginia Woolf's unique ability to aesthetically crystallize moments in writing stems from how she magnifies rather than neutralizes the risk of finitude that grounds experience (2012: 56–78). The aesthetic and affective resonance of the novels Hägglund explores resides not in "an intimation of eternity," but, on the contrary, in the palpable "investment in a life that is susceptible to transformation and loss" (2012: 58)—the drama of these works arises from an affirmation and intensification rather than a negation of transience. What persistently emerges from Hägglund's literary analyses is that even in the novels' most life-affirming of moments—*especially* in the novels' most life-affirming of moments—there is always that lingering sense that *this can never last*. Only against the horizon of time and finitude can the arcadian appear.

One must be careful here not to extend existential theorizations too readily to empirical reality, because chronolibido, like the existential primacy of imperfection, is in itself insufficient to account for specific aesthetic or ethical positions. The fact that loss and desire are intertwined is not enough to guarantee that one will care for a certain object or be predisposed to a particular existential attitude. Hägglund's chronolibidinal readings of literature nonetheless demonstrate how an object can derive its emotional allure from appealing to our faculties of chronophilia and chronophobia. By aesthetically configuring a particular dynamic between these two faculties, an art object can elicit passionate feelings—as, for example, when loss is staged as imminent (Clune and Hägglund 2015: 120) or when the sense of time's passage is aesthetically amplified (Hägglund 2012: 45). While chronolibido primarily delineates a fundamental perceptual condition with no given aim, Hägglund thus also invokes, but never fully specifies, a notion of "chronolibidinal aesthetics" (2012: 19) to gesture at the power of aesthetics to mediate, magnify, and dramatize the

---

[3] Also of note here is literary scholar Virgil Nemoianu's *Imperfection and Defeat* (2006). In this work, Nemoianu, while not explicitly formulating a logical bond between loss and desire, conceptualizes literature as a privileged medium for elucidating the centrality of loss and imperfection to existence.

workings of the temporal. Such chronolibidinal aestheticizations, I propose, often fall under the aesthetic rubric of imperfection, indicative as imperfection is of the corrosive effects of temporality.

A prominent example of such a chronolibidinal aesthetic of imperfection is found in American composer William Basinski's famed *The Disintegration Loops*, a 2002–3 collection of albums whose aesthetic of imperfection is premised on technological and sonic collapse. Basinski serendipitously discovered the sound for these records when he attempted to transfer part of his archive of analog tape loops to a digital format. As Basinski ran his loops through a digital tape recorder, he noticed that the tape gradually deteriorated in its physical contact with the tape head. As a result of this, the music audibly disintegrated as the tape simultaneously inscribed and erased itself. I have elsewhere argued that this principle of "death by inscription" performs a material enactment of autoimmunity: within the technical recording assemblage, the loops could survive as a trace only by subjecting themselves to the violent alterity of the tape recorder (Kemper 2019). The resultant melancholic symphonies of decay, released by Basinski as *The Disintegration Loops*, were met with renown precisely for their slowly self-annulling sound. Their widespread resonance can largely be explained by conceiving of them as a profound form of chronolibidinal aesthetics. By aesthetically communicating their demise to the listener, the loops appeal to the faculty of chronophobia, or the fear of loss, and thereby amplify the feeling of chronophilia, or the intensity of one's investment in a sound that is audibly under duress.[4] The pathos of the loops stems, in other words, from how their aesthetic presentation addresses our capacity to understand what it means to lose and to disintegrate.[5]

---

[4] This sense of loss is conceptually redoubled by the record's ties to 9/11. Basinski, who witnessed and recorded the day's cataclysmic events from his Brooklyn rooftop, released *The Disintegration Loops* in the aftermath, set to footage of the disfigured New York skyline. This caused the music to become something of an "unofficial elegy" to 9/11's tragedies (Richardson 2012).

[5] As remarked in the Introduction, aesthetic appreciations of finitude are certainly not always noble or commendable. For example, while one could analyze the phenomenon of ruin porn—the widespread aesthetic appreciation of (footage of) dilapidated (urban) environments (Lyons 2018)—through the lens of chronolibido and explain the appeal of ruined architecture by pointing to its visualization of time's passage, such a perspective would risk neglecting the ethical implications of appreciating signs of destitution. The city of Detroit, for instance, is a popular site of ruin photography, but its sprawls of urban decay signify decades of poverty that are generally elided in photographers' stylized shots (cf. Gansky 2014; Wells 2018). Gaining aesthetic pleasure from traces of imperfection and disrepair can problematically gloss over real suffering and hardship (Saito 2017b). Chronolibidinal attachments to objects that appear fragile are thus not *a priori* ethically admirable and one should heed against fetishizations of finitude that sanitize or romanticize the pain and misery that run-down scenes so often symbolize.

The notion of chronolibidinal aesthetics underlines the relation between imperfection as an existential condition, on the one hand, and an aesthetic category, on the other: while the existential primacy of imperfection in itself precipitates no particular aesthetic orientation, objects that exhibit an aesthetic of imperfection can arouse in their audience, through their representation of processes of loss and erasure, a chronolibidinal sense of imperfection's constitutive workings. In what follows, I will primarily be concerned with how the flow of chronolibido can be affected through a *technological* aesthetic of imperfection and with the forms of care that this can generate. In Chapters 3 to 5, I will analyze several technological objects that, like *The Disintegration Loops*, make tangible the bonds between loss, desire, and finitude. These objects, in their own way, encourage a form of care for the technological realm by composing a chronolibidinal aesthetic of imperfection. However, in order to better understand the relation between imperfection, aesthetics, and technology on which these objects rely, it is apposite to first shift attention to another central concept that guides my study, namely the Derridean notion of *spectrality*. This concept is inherent to how the existential primacy of imperfection takes shape, but also offers an ethically and politically charged lens through which to analyze the aesthetic proclivities of today's technologies, a claim that I further develop in Chapter 2.

Spectrality

There have always been spectral shapes flickering in the margins of Derrida's work, but it is in his 1994 text *Specters of Marx* (originally published in French in 1993) that he deals most intently with ghostly affairs. In this text, Derrida develops his now canonical notion of hauntology. Hauntology stages a challenge to ontology's implications of fixity, essence, purity, and presence: "[I]t is necessary to introduce haunting into the very construction of a concept. Of every concept, beginning with the concepts of being and time. That is what we would be calling here a hauntology. Ontology opposes it only in a movement of exorcism" (Derrida 1994: 202). Derrida takes up this spectral concept in a number of ways, but one can already glean from this passage that hauntology is, for him, principally a constitutive condition that describes how a relation of haunting, like autoimmunity and imperfection, informs all that exists. Hauntology and its figure of the ghost effectuate a necessary blurring of boundaries, tainting the rigidity of distinctions between, for instance, presence and absence, life and death, past and future. Instead of holding these pairings to be binary oppositions, hauntology reveals them to work through one another, making it so that a concept is never free from being haunted by what is other than itself. Hauntology and the related notion of spectrality[6] invalidate any chance of purity and perfection; the possibility of the ghost is precisely the possibility of contamination that defines Derrida's deconstructive procedures. Accordingly, in Derrida's work, ghosts serve not as paranormal emissaries but as symptoms of the ultimate impossibility of "clean" or perfect ontologies and teleologies.

A recurring motif in *Specters of Marx*, and, in fact, the prime reason that we can think of such a thing as the ghost, is the disjointed nature of time (Hamlet's famous exclamation that "the time is out of joint" is one of the themes around which Derrida's text finds articulation). A specter or a ghost is something that returns from the past and/or arrives from the future— in any event, it erodes whatever fantasies of self-containment are grafted onto the present and thereby "recalls us to anachrony" (Derrida 1994: 6). This is illustrated by the following passage, where Derrida elucidates the anachronism that is endemic to time itself:

> If there is something like spectrality, there are reasons to doubt this reassuring order of presents and, especially, the border between the present, the actual or present reality of the present, and everything that can be opposed to it: absence, non-presence, inactuality, virtuality, or even the simulacrum in general, and so forth. There is first of all the *doubtful contemporaneity of the present to itself*. Before knowing whether one can differentiate between the specter of the past and the specter of the future, of the past present and the future present, one must perhaps ask oneself whether the *spectrality effect* does not consist in undoing this opposition, or even this dialectic, between actual, effective presence and its other. (1994: 48, emphasis added)

If we recall that the spacing of time ensures that each moment is divided within itself, it becomes clearer what Derrida means when he speaks of the "doubtful contemporaneity of the present to itself." That is, before we can even assess whether it is the past or the future that has come to

---

[6] There are certainly differences between these two terms (hauntology is a specifically Derridean concept while spectrality comprises a broader discourse), but, like Derrida, I will use the terms largely interchangeably, because my usage can always be tied back to Derrida's thinking on autoimmunity and the trace. Similarly, while important differences exist between the ghost and the specter (the latter carrying overtones of visibility and surveillance, and the former harboring more pronounced associations of vulnerability and temporality), I again use the terms relatively interchangeably. For a more in-depth elaboration of the differences between specters and ghosts, see Fisken (2011); Mladek and Edmondson (2009).

assert itself in the present, that very present is already disjointed in its becoming. Temporal succession does not describe a sequence of insulated moments, but rather a process of persistent self-negation—each moment is simultaneously present and absent in the sense that it elapses in the very instant that it becomes and is thereby necessarily bound to a past and a future through the movement of spacing and tracing. The present's non-contemporaneity in turn opens up the material conditions by virtue of which past and future may always manifest themselves, sometimes in unexpected ways.

The concepts of hauntology and spectrality thus accommodate temporality's twofold disjointedness. First, these concepts underline the logic of spacing, revealing how the division that penetrates the present necessitates a foundational process of haunting. Second, spectrality and hauntology register the materially situated ghostly effects that this disjointedness makes possible, allowing historically and culturally specific specters to emerge (this is a dimension I will return to in the following chapter). As a material consequence of spectrality, there are many different ways of inhabiting time—of living in relation to the residues of the past and the contingencies of the future—and imposed temporalities, no matter how inflexible they seem, may always be contested. Artist and theorist Rasheedah Phillips describes, for example, how "[a]lternative temporalities embodied by such cultural movements as Afrofuturism, and DIY theories as Black Quantum Futurism (BQF), have developed practical tools and technologies for exploring reality and shaping past and future narratives" (2020: 242). There is, then, a political dimension to the concepts of hauntology and spectrality, as they can help to challenge culturally and socially construed illusions of fixed temporal trajectories.

The central concepts I have thus far been drawing on all disclose a ghostly logic. Autoimmunity, to reiterate, designates that all things carry the germs of their own dissolution as a result of the spacing of time. Spacing spells out that the structure of haunting is universal: one cannot prevent the frequentation of ghosts, both as harbingers of death and as figures of difference, because of the internal and irreducible exposure to alterity that is wrought by temporality. In its simplest definition, haunting describes the impossibility of temporal self-containment and the attendant plight that a trace, in order to exist at all, must be placed in a contaminative relation to a past that constrains it and to a future that may efface it. The promise of death haunts existence from its inception; existence cannot be what it is without incorporating the spectral and violent threat of finitude. Seen in this light, philosopher Giorgio Agamben's description of spectrality as "a form of life" that is "perfect, since it no longer has anything to add to what it has said or done" (2013: 475) is puzzling. The disjointed nature of time ensures that the past can never be fully reckoned with and that the future remains uncertain, and the resultant, corruptive condition of spectrality directly contradicts Agamben's invocation of perfection. Because of the ultimate impossibility of a perfect and self-contained integrity, no being can be exempt from the *im*perfect workings of the spectral, insofar as this describes an unassailable process of contamination and finitude.

Chronolibido concerns the human internalization of this overarching condition of spectrality and imperfection. The concept of chronolibido defines human time-consciousness as premised on the recognition of an indelible logic of finitude. Chronolibido is, therefore, a thoroughly spectral notion; it is no incident that in *Dying for Time*—his prime work on the concept—Hägglund uses the term "haunting," or derivations thereof, as often as thirty-five times. The ghostly nature of chronolibido is, however, captured even better by Derrida's previously recounted commixture of love and ruin. As we saw, in Derrida's view, one can love something only through "the *ghost* or the silhouette of its ruin, one's own ruin—which it already is, therefore, or already prefigures" (2002: 278, emphasis added). The ghostly prefiguration of death that circumscribes both the object of one's love and the experience of love itself is a requisite element of the formation of care and desire. As Hägglund's concept of chronolibido clarifies, care can emerge only if loss is intuited as a possibility. All libidinal investments depend, in other words, on how an anticipation of erasure already suffuses the moment. It is only due to a mode of haunting, due to the premonition of a future ruin that roots the present, that any affective investment can be elicited.

Most of the claims that have thus far been set forth are highly general, describing the *a priori* conditions of existence rather than establishing the way these conditions play out materially. What, then, of my more specific interest in a technological aesthetic of imperfection as it unveils and orients people toward these existential constraints? In addressing the technological aspects of this question, spectrality has a primary role to play. Here, it is important to explicitly distinguish different layers of my analysis so as not to conflate them. So far, I have presented spectrality primarily as a constitutive and therefore ahistorical condition, in the sense that its logic is not tied to a specific material context. Yet, this foundational condition of spectrality enables the possible manifestation of historically, culturally, and technologically specific ghosts. This describes a material rather than an existentially undifferentiated level of analysis: spectrality allows one, through the figure of the ghost, to conceptualize how past and future, life and death, the visible and the invisible are materially and empirically active in a historically and technologically situated present, and this grants it an explanatory value when analyzing technological processes. To be sure, no ghostly manifestation can have any direct bearing on the broader logic of spectrality: no matter how many ghosts we welcome or exorcize, we can never dispel this overarching condition. Yet, ghosts, through the specificity of their apparition, allow one to better understand the material, technologically mediated effects of the unassailable logics of finitude and imperfection. However, to fully develop this argument, as I will do in the following chapter, it is necessary to briefly bracket the question of the spectral and to first assess the constitutive effect of technology on the human (chronolibidinally charged) recognition of the temporal.

Postphenomenology and the *pharmakon*

The question of how technology impacts our perception of and relation to the world and its existential constraints has been central to the rising field of postphenomenology (Ihde 1990; Rosenberger and Verbeek 2015). Postphenomenology, most prominently associated with philosopher Don Ihde, maps how human beings relate to their surroundings through the necessary mediation of technology. More concretely, postphenomenological perspectives "investigate technology in terms of the relations between human beings and technological artifacts, focusing on the various ways in which technologies help to shape relations between human beings and the world. They do not approach technologies as merely functional and instrumental objects, but as mediators of human experiences and practices" (Rosenberger and Verbeek 2015: 9). As opposed to phenomenological traditions that view technologies as mere objects in the world or even as having an alienating effect on the purported purity of the human subject, postphenomenology stresses that technologies actively shape the sense that human beings have of the world; rather than claiming that technology perceptually binds an already-formed subject to an already-given world, postphenomenology conceives of technology as elemental to the relation between both (Rosenberger and Verbeek 2015: 11–12). Accordingly, postphenomenology "analyzes the character of the relation human beings have with . . . technology and the ways in which it organizes relations between human beings and the world. . . . Technologies, to be short, are not opposed to human existence; they are its very medium" (Rosenberger and Verbeek 2015: 12). In developing the arguments in this section and in my later case studies, I have been inspired by the postphenomenological approach, particularly by how it conceives of technology as something that is inseparable from aesthetic and ethical evaluations of the world. Philosophers Robert Rosenberger and

Peter-Paul Verbeek describe the defining qualities of postphenomenological perspectives as follows:

> First of all, they typically focus on *understanding the roles that technologies play in the relations between humans and the world, and on analyzing the implications of these roles*. . . . This focus on human-technology relations implies, second, that *postphenomenological studies always include empirical work as a basis for philosophical reflection*. . . . Third, *postphenomenological studies typically investigate how, in the relations that arise around a technology, a specific "world" is constituted, as well as a specific "subject."* . . . Fourth, on the basis of these three *elements, postphenomenological studies typically make a conceptual analysis of the implications of technologies* for one or more specific dimensions of human-world relations—which can be epistemological, political, aesthetic, ethical, metaphysical, et cetera. (2015: 31, emphasis original)

It is particularly this question of how technology modulates the aesthetic and ethical dimensions of how human beings feel out and make sense of the world that has my interest, considering that this question bears directly on the existential notions of finitude and imperfection I explore.

Concretely, following the premises of postphenomenology, one may surmise that, while autoimmunity is a foundational condition that precedes the human—"more primordial than what is phenomenologically primordial" (Derrida 1973: 67)—its human phenomenologization and the conjunctive flowing of chronolibido are co-composed through technology. The work of philosopher (and former student of Derrida) Bernard Stiegler helps to unpack the postphenomenological assertion that technology has a privileged function when it comes to exposing and attuning people to the existential conditions I have so far discussed. To be sure, the postphenomenological claim about technology does not negate Hägglund's argument that chronolibido is in itself insufficient to compel a particular disposition. Rather, it suggests that technology is a constitutive factor in *how* the temporal is perceived and aesthetically experienced, and that technologies preform pathways for the direction, intensification, and diminution of care. By implication, the technical dimension of temporal experience impels a need, in contrast to the relatively apolitical character of Hägglund's work on chronolibido,[7] to "politicize phenomenological questions" (Stiegler 2019:

---

[7] Hägglund's more recent work *This Life* (2019) does develop a more politically charged perspective on finitude, but he ties this predominantly to Marxian economics and not to the field of aesthetics.

76) and to critically analyze how technical constellations comprise the conditions of possibility for care to emerge. As the example of Basinski's *The Disintegration Loops* evinced, a technological aesthetic of imperfection can be conducive to chronolibido's tidings. Stiegler's work helps to unravel the political, ethical and critical dimensions of such an aesthetic, especially vis-à-vis the contemporary technological climate.

The overarching assertion that guides Stiegler's work is that the human relation to the temporal

> is always already determined by its techno-logical, historical conditions, effects of an originally techno-logical condition. Time is each time the singularity of a relation to the end that is woven technologically. Every epoch is characterized by the technical conditions of actual access to the already-there that constitute it as an epoch, as both suspension and continuation, and that harbor its particular possibilities of "differ*antiation*" and individuation. (1998: 236, emphasis in original)

For Stiegler, technology and human consciousness are indissociably intertwined. Humans are and always have been technological beings through and through; the human brain, with its marked capacity to apprehend time, is formed by its engagement with and dependence on technical traces.[8] Stiegler expands Edmund Husserl's classic model of time-consciousness by supplementing Husserl's notions of *primary retention* (direct temporal perception) and *secondary retention* (memory and imagination) with a necessary third, exteriorized form that affects the former's interplay: *tertiary retention*, or external and artificial memory (Stiegler 2011a: 39).[9] Technical objects materialize time in a particular fashion and thereby also impact and restrict one's temporal perception. The way in which Stiegler repurposes Husserl's example of the melody is informative here. Whereas it was once "impossible to hear the *same* melody twice," the invention of recording technologies makes it possible to play and replay a singular temporal object, which has far-reaching implications for how temporal experience,

---

[8]   It might seem as if Stiegler proposes a binary distinction between the technical human and the non-technical, purely instinct-driven animal. This is, however, not how Stiegler's account of the human should be read; humanity, for him, is "a behavior potentially performable by all kinds of life" (Moore 2013: 27). This behavior can be summarized as the externalization of consciousness and memory in technical traces (exosomatization), which thereby opens up a perception of time along with the possibility of technical rather than merely genetic evolution.

[9]   The concept of tertiary retention will be particularly important for Chapter 5's analysis of the musical genre of vaporwave.

in its dependence on retentional processes, is shaped (Stiegler 2009a: 54, emphasis in original). That the melody can now be inscribed as a repeatable, shareable, and industrially exploitable, technical trace changes one's potential perception of it, in turn affecting the selection criteria through which secondary retention (memory and imagination) informs primary retention (direct temporal perception). It makes a difference whether I hear a song for the first time or for the tenth time, and technology regulates possibilities for mediation and repetition.

Temporal experience is thus (in contradistinction to Husserl's model, where the composition of temporal experience is a wholly internal process) partially formed through extraneous technical objects. The dependency on exterior artifacts ensures that any sense of human subjectivity and interiority is always modulated by a technological outside: "the *who*," Stiegler frequently repeats, "is nothing without the *what*" (1998: 141, emphasis in original). This recursive dependency is, moreover, underpinned by the general conditions of autoimmunity and imperfection. Consciousness itself already describes a fragile apparatus whose retentional capacities are finite and susceptible to slippages. Yet, autoimmunity also ensures that there exists an irreparable "fragility of the technical element" (Stiegler 2019: 13). This fragility, I posit, operates on two dimensions. First, technical artifacts are themselves finite: there is no technology that could circumvent the spacing of time to nullify autoimmunity and the risk of destruction. Second, a mind's relation to its technical prostheses is never given once and for all, and consciousness is always open to being reformed by the technical milieu in which it exists (Stiegler 1998: 158). This claim aligns with the postphenomenological notion that technologies shape an open field of relations in which subject and object, insofar as it is even justified to speak in these terms, are mutually constituted. By implication, the heterogeneity of the technical realm spells the need to think carefully about how technology impacts perception and about the forms of investment this facilitates or impedes, especially if we want to remain responsible for the multifarious effects our technologies reap.

The fundamentally ambiguous effect of technology on perception is best captured by Stiegler's mobilization of the classic philosophical notion of the *pharmakon*, a concept that will remain a central thread throughout this study (see also Kemper 2022). Stiegler inherits this concept from Derrida, who in turn traces it back to Plato's *The Phaedrus*. In *The Phaedrus* (2002), Plato stages a dialogue between Socrates and Phaedrus that unfolds during a rest on the banks of the river Ilsios. Part of the dialogue consists of a recitation by Socrates of the myth of the Egyptian god Theuth who brought the gift of writing to King Thamus. This recital is intended to point out a certain double logic to how the technology of writing impacts the human mind. According to Socrates, Thamus rejected the gift that Theuth had brought to him, as he believed that this purportedly beneficent technology would turn out to bring adverse effects. Thamus stresses that writing, as a technical process that allegedly offers an unprecedented aid to memory, in fact diminishes one's own ability to memorize. The more one relies on writing and paper to do one's memorizing for one, the more one actively curbs the mind's capacity; "under pretext of supplementing memory, writing makes one even more forgetful; far from increasing knowledge, it diminishes it" (Derrida 1981a: 100).

Plato, Derrida posits, thus presents writing as a *pharmakon*: something that is both poison and cure. Yet, what Plato, according to Derrida, did not sufficiently recognize is that, within the structure of the *pharmakon*, the good and the bad are by default interwoven; toxin and tonic are produced in the very same technical gesture (1981a: 103). It thus does not make sense to speak of technology as all good or all bad; technologies, in augmenting or supplementing certain cognitive and sensory capacities, necessarily dull or diminish others.[10] Derrida, moreover, criticizes Plato's presentation of writing as an artificial element that dilutes the supposed purity of the human mind (1981a: 110). The *pharmakon* of writing is not an external object that impinges on human consciousness, but rather indicative of a formative relation between humans and the world.

We are, as established through the work of Hägglund, chronolibidinal beings, but Stiegler expands on Derrida's claims to demonstrate that we are also *pharmacological* beings (Stiegler 2013: 43, emphasis added): beings whose perceptual, sensory, cognitive, and affective capacities are both expandable and diminishable by technology. For Stiegler, the pharmacological relation is not exclusive to writing but structures all technological mediations; as his concept of tertiary retention discloses, technologies always have the capacity, for better and for worse, to affect how human beings experience the world. There exists no pre-technological purity: the mind is by default dependent on technological *pharmaka* that always carry the potential both to enlarge and to impair, both to intensify and stifle one's relation to the world (Stiegler 2013: 25).[11] Pharmacological thinking thus embraces the notion that technology always co-constitutes consciousness, that one's relation to technology is never definitive and that technology therefore may always bring both prosperous

---

[10] The concept of the *pharmakon* bears similarity to the notion of "tradeoffs" that is prevalent within postphenomenological discussions. As Rosenberger and Verbeek describe, "through the non-neutral transformations rendered to user experience through the mediation of a technology, we not only receive the desired change in our abilities, but always also receive other changes, some of them taking on the quality of 'tradeoffs,' a decrease of a sense, or area of focus, or layer of context" (2015: 16).

[11] On the concept of the *pharmakon*, see also Pisters (2021).

and calamitous effects. One crucial consequence of this situation is that technological *pharmaka* enact a double role: because of their mediating role in experience and the attendant formation of care and desire, *pharmaka* should themselves be *carefully* considered and managed to regulate whatever poisons they may hold. As Stiegler contends, "[t]he *pharmakon* is at once what *enables* care to be taken and that of *which* care must be taken—in the sense that it is necessary *to pay attention*: its power is *curative to the immeasurable extent* that it is also *destructive*" (Stiegler 2013: 4, emphasis original). Pharmacological approaches should thus map how *pharmaka*'s curative and poisonous capacities manifest themselves at least partially through how a society or community relates to and cares for its technologies. Significantly, they should thereby also investigate the kinds of chronolibidinal relations to the world and its material conditions that technology shapes or inhibits.

In fact, in his readings of Nabokov, Hägglund unwittingly comes close to teasing out the pertinence of pharmacology in relation to chronolibido, although he does not follow this argument through. What Hägglund's inquiries primarily show is that Nabokov was so saturated with chronophilia— so extraordinarily attuned to the weight of every moment, so preternaturally attached to the fragile world around him—that his entire life was regimented by the chronophobic project of preserving, through the technicity of writing, as much of his universe's transitory beauty as he humanly could. As Plato already recognized, writing is never a neutral practice. Rather, it works its way back on the practitioner, in Nabokov's case further sharpening the acuity of his perception (cf. Hägglund 2012: 84). Nabokov's 1969 opus *Ada or Ardor* most directly examines the dramatic act of exteriorizing a life in technical traces. While Ada and Van, the novel's amorous protagonists, have remarkably strong skills of memory, they are also, as a prime empirical example of Stiegler's concept of tertiary retention, "dependent on supplementary devices to retain the flight of time" (Hägglund 2012: 98). The novel is pervaded by technologies that are employed as chronolibidinal conduits that facilitate remembrance or help maintain the moment, but that are simultaneously shown to be just as corruptible and fallible as the human mind itself (the centerpiece of this dynamic is the memoir the protagonists are composing, which remains necessarily fragmented and incomplete, and is later modified by a jealous editor). Following Hägglund, the novel's focus on the fragile materiality of technology reveals temporal perception to be always contingent on the precarious mediation of technical traces—traces that constantly need to be invested in and attended to (2012: 99, 104). Hägglund's analysis, then, implicitly confirms Stiegler's postphenomenological perspective, as it suggests that temporal experience and chronolibido crystallize through externalized, technological processes of mediation.

However, the enmeshment of care and technology reveals there to be one major shortcoming to Stiegler's (and, by implication, Derrida's) conceptualization of the *pharmakon*: Stiegler largely neglects the question of how technology connects us to the other, and particularly those others directly implicated in the composition of one's pharmacological experience. Stiegler, by focusing so intently on how the individual mind is pathologically affected by its technological milieu (predominantly grieving the contemporary spread of stress, anomie, and depression), de-emphasizes the question of how the mind is alerted to, and itself affects, the other lives and objectual finitudes that are at stake in the *pharmakon*. Care for one's mind through one's engagement of the *pharmakon* may always come at the expense of the lives of others—as such, any pharmacology that only heeds the poison that pollutes the user's individual mind is inadequate.[12] Pharmacological thinking must take an ecological view that recognizes that an individual's ostensibly salutary relation to technology is often directly facilitated by the pernicious struggles of others and that technologies may, to various degrees, restrict or enable a recognition of this situation. Technological *pharmaka* are, moreover, themselves material agents that, in their manufacture and disposal, produce literally poisonous effects.

These reflections become especially pressing in view of today's technological situation, in which the supposedly smooth user experiences that digital technology affords can only be realized through the simultaneous extension and obfuscation of a vast network of exploited laborers, extracted resources, and expended energy. This describes a pharmacological state in which there is a (purportedly) capacitating augmentation of perception in the form of instant connectivity and frictionless user experience while, in the same stroke, there is a diminishment of the user's ability to apprehend the material conditions of technological production and consumption. While this argument will be fully fleshed out in Chapter 2, it is important to already gesture toward it here, because it shows the importance of understanding the experiences that *pharmaka* shape in relation to the networks that sustain that experience.

---

[12] A related weakness in Stiegler's pharmacological project is that he frequently presents the *pharmakon* in a totalizing light, as a technical assemblage that affects all its users in a roughly evenhanded fashion. He thereby tends to neglect the variegated ways in which technology impacts different segments of the populace and fails to account for how this differentiated impact is animated by intersecting socioeconomic, political, and cultural conditions. The generalized "we" that Stiegler commonly invokes (see, for instance, Stiegler 2017b: 387) when discussing the impact of today's *pharmaka* is thus partial; technology is never the only factor to shape the mind, and one's access to, engagement with and aspirations for technology are always informed by a host of contextual aspects—aspects that intersect with matters of class, race, gender, and location. See, for example, André Brock Jr.'s notion of critical technocultural discourse analysis (CTDA) for an approach that does more justice to the situated aspects of technology use (2020: 2).

Restricting the interpretation of poisons and cures only to the limited effect a pharmacologically enabled experience has on the individual user's own life, without considering what else is at stake in that experience, will fail to engender a truly sustainable and collective pharmacology. Thinking in terms of the *pharmakon*, then, should raise questions about how a technology does or does not encourage users to assume responsibility for the lives and deaths at play within that technology. Pharmacological approaches must, in other words, acknowledge that the incapacitating effect of a technology might consist in how technology fails to attune users to the material conditions of its production and should assess how the user's experience of a technology is entangled with the existence of those that are tasked with facilitating that experience. Such perspectives should also stress that the potentialities of the *pharmakon* are never exhausted; there are always ways of retuning its curative and poisonous capacities, for example through exploring aesthetic reconfigurations of the interplay between technology, human beings, and the world.

Taking seriously Stiegler's claim that the *pharmakon* "*enables* care to be taken" but is also "that of *which* care must be taken" (2013: 4, emphasis in original), I will, in the following chapters, approach technology as something that can, through its aesthetic properties, *itself* become an object of chronolibido. The objects that comprise my case studies—especially the artworks I analyze in Chapters 3 and 4—draw on an aesthetic of imperfection to encourage an emotional investment in the technological realm. These investments are significant for how they link the pharmacological to the chronolibidinal, revealing that the latter can be mobilized as an aesthetic concept with political, ecological, and environmental implications. The pharmacological aspects of chronolibido pertain to how different aesthetic figurations of technology enable different relations to technology's finite and fragile conditions of production, consumption, and disposal, which may also spark a more serious involvement with all that hangs suspended in the opaque webs technology weaves. I thus approach the *pharmakon* principally in an aesthetic register, questioning how a technological aesthetic of imperfection, through its intimation of the temporal, may evoke more curative relations to technology.

Humans, Technology, and the Existential Primacy of Imperfection

The notion of pharmacology underlines why it is important that we, as humans, remain cognizant of the forms technology assumes. Here, I again take cues from the perspective of postphenomenology. While sympathetic to philosophies that destabilize the centrality and sovereignty of the human, "[t]he postphenomenological approach . . . explicitly does not give up the distinction between human and nonhuman entities . . . [because] [w]hen we give up this distinction, we also give up the phenomenological possibility to articulate (technologically mediated) experiences 'from within'" (Rosenberger and Verbeek 2015: 19–20). To relinquish the possibility of articulating human experience from within would, I contend, also risk absolving the human from its responsibilities toward its surroundings. Without giving recourse to anthropocentrism or human exceptionalism, we must accentuate the peculiar position the human occupies as a technological agent that can consciously take on obligations toward the world.

Philosopher Franco "Bifo" Berardi most forcefully conveys the significance of this claim. While few theorists have better documented how today's *pharmaka* of speed and connectivity produce psychopathological dismay in the planetary form of apathy and anxiety, here I am more interested in Berardi's musings on the status of the human vis-à-vis the technological. Berardi offers a remarkable meditation on whether the fragile human body could ever be emulated by a technological automaton and it is around his words that the scope of my study coalesces:

> My question is not about the technical feasibility of a perfect simulacrum, but about the emergence of the self, by which I mean the self-perception of a conscious singularity. The *sense of duration* is the essential mark of the conscious self, and the irreducibility of existence to algorithmic recombination is based on this. The sense of duration cannot be simulated in an artificial construct that perfectly reproduces the features of the human body, because *the sense of duration is not behavior, but suffering, consciousness of the organism's decomposition, and consciousness of death.* (2015: 288, emphasis added)

What Berardi's otherwise trenchant assertion[13] does not explicate is that the consciousness of death and decomposition that distinguishes us from our machines does not merely grant us a sense of our own finitude; rather, it suggests that we also recognize the finitude and fragility of others. What is,

---

[13] Berardi's argument resonates with other recent discussions of the qualitative differences between the human and its digital machines. Such reflections are found, for example, in philosopher Kwame Anthony Appiah's conceptualization of our inescapable constitution as "imperfect creatures, with imperfect memories and capacities for calculation" (2017: 44)—something that ensures that our understanding of the world is necessarily fractured, but all the richer for it (2017: 110). Likewise, computer scientist Melanie Mitchell's recent study of the current state of Artificial Intelligence suggests that what differentiates our own intelligence from the intelligence instantiated in AI is not so much the ability to create or to perform complex cognitive tasks as it is the ability to find meaning in what we do and in the objects that surround us (2019: 274).

then, unequivocally central to what it means to be human is that we *intuit the existential primacy of imperfection*: we display a pronounced awareness of the autoimmune nature of existence and can comprehend finitude as a collective and immanent condition. We have, moreover, a *fundamental capacity for forming chronolibidinal attachments*: we can become invested in and care for things *because* we recognize their fragile nature. By implication, we can appreciate chronolibidinal aesthetics: as beings that are conscious of the cast of mortality, we can be stimulated by aesthetic expressions of loss and ephemerality. Lastly, we can *purposely form a relation to spectrality* both as a universal condition and as the always contextual appearance of ghosts: we can inhabit the temporal with a sensitivity to its corrosive effects and can choose to be responsive to specific ghostly manifestations. The faculty of recognizing the constitutive primacy of imperfection and the capacity of care this engenders unveil a human condition that is, at least as yet, irreconcilable with the technological tendencies that perhaps outstrip us in other regards. The chronolibidinal investments to which our sense of spacing and the attendant necessity of mortality give way mark a crucial trait that our machines have not yet mastered (and perhaps never will), and they form the basis on which we give our lives, but also our technologies, their meaning.

In contrast to the relatively apolitical character of his earlier works, Hägglund has recently elucidated the ethical demand for justice that attends an earthly coexistence defined by collective finitude. He asserts that, because no being is ever entirely self-sustained and because finitude necessarily defines not only our own life but also (our relation to) the world around us and its many inhabitants, a sense of the fragile composition of the world is precisely what renders our actions meaningful and what should urge us to take care of our surroundings:

> [T]he peril of death is an intrinsic part of why it matters what we do and why it matters that we devote ourselves to someone or something living on beyond ourselves. We have to take care of one another because we can die, we have to fight for what we believe in because it lives only through our sustained effort, and we have to be concerned with what will be passed on to coming generations because the future is not certain. (Hägglund 2019: 168)[14]

---

[14] Here, Hägglund's work resonates with that of Stiegler, as the latter criticizes Martin Heidegger's notion of being-towards-death for how it fails to "investigate the protention of what happens beyond the end of Dasein," thereby omitting the important capacity of care to be magnetized by what transcends and outlasts one's own life (2019: 259).

While Hägglund implies that the collective awareness of death stipulates a set of moral injunctions, a sense of shared mortality in reality seldom proves sufficient for the construction of a more equitable world. As the ongoing desecration of the planet makes plain, the mere fact that human perception is informed by an apprehension of mutual finitude is certainly no guarantee for ethical and inclusive ways of living, or for a desire for technological forms of conduct that diminish rather than increase destruction. Hägglund's compelling account shows that making explicit life's existential constraints enables us to think more carefully about the world we build, but the collectivity of finitude in itself provides little assurance or direction for practical relations of care.

Based on the aforementioned, it makes sense to view our intuition of the existential primacy of imperfection as a necessary starting point but never as the final node in developing ethical perspectives or more sustainable interactions with technology. It is important to stress that the capacity to recognize finitude makes us uniquely answerable for how the conditions of imperfection, autoimmunity, and spectrality materially play out. Accordingly, if we nowadays delegate an ever-greater degree of tasks to technology, it is all the more crucial to question the material consequences of our technological dependencies. The responsibility of caring for a dying planet and its vulnerable denizens, driven as this labor needs to be by a feeling for what it means to lose and pass away, cannot be given over to our machines. When we transfer our embodied knowledge to our technologies, we risk excising how human embodiment uniquely affects that knowledge: it grants the ability to intervene in processes on the basis of an awareness of the existential primacy of imperfection and its material implications.[15] Within the multitude of practices that come with building a world, it is thus vital to heed against allocating a disproportionate amount of agency to technology— not out of conservative convictions, but out of a pharmacological sensibility that recognizes that technology has no sense of and concern for its own toxicity. It is incumbent on us, as humans, to let our apprehension of finitude and fragility inflect how technologies unfold by reining in any poisonous excesses.

---

[15] To reiterate, my aim here is not to unfold a narrative of human exceptionalism, but to show the serious ethical obligation that our recognition of the primacy of imperfection bestows upon us as humans, without wanting to deny that other animals may betray similar traits. Our sense of finitude is, moreover, not limited to human bodies, and thinking more meaningfully about mortality would ideally mitigate what environmental philosopher Thom van Dooren describes as today's widespread "inability to be *affected* by the incredible loss of this period of extinctions, and so to mourn the ongoing deaths of species" (2014: 18, emphasis in original).

As the previous section underlined, technological *pharmaka* themselves play a key role in how this apprehension takes shape. As chronolibidinal and pharmacological beings, humans can assume responsibility for but are at the same time deeply moldable by their technologies. This foregrounds what I have presented as the most pressing function of pharmacological thinking: it must analyze how technology materially organizes conditions of finitude and fragility *and* how these conditions are, or are not, perceptually, aesthetically, and affectively communicated to the user. If consciousness is indeed composed through technology, then any reflection on one's modes of thinking and living must involve the question of how technological *pharmaka* are mobilized and how such *pharmaka* mediate one's world (cf. Stiegler 2019: 224). The analyses I undertake in Chapters 2 to 5 underline how technological aesthetics premised on imperfection can substantially shape the direction, scope, and intensity of investments, giving a practical, aesthetic edge to the concepts of autoimmunity, chronolibido, and spectrality by foregrounding their pertinence to pharmacological questions.

In the chapters that follow, I will regularly engage work done in the fields of media studies, the environmental humanities, and the ethics of care. At face value, these more materially oriented inquiries perhaps appear incongruent with the ostensibly a-prescriptive, post-Levinasian ethical framework of Derrida and Hägglund. To be sure, it is true that the central concepts that I have outlined do not in themselves shape moral values and that no unconditional principles or ideals can be drawn from them: not everything that is finite should be protected, not everything that is fragile should be cared for. Yet, this does not legitimate a moral relativism—on the contrary, the deeply material process of spacing is what enjoins a contemplation of phenomena in their situated, ghost-producing specificity. The impossibility of a transcendental ethical axiom or culminating perfection is precisely what necessitates the analysis of finitude and imperfection as always also materially organizable phenomena. Moreover, the fact that Derrida issues no categorical injunctions is precisely what should urge us to consider the contextuality of care. Accordingly, the more practical and morally charged arguments that will be elaborated in the rest of this book should not be seen as a challenge to but rather as an extension of the material project Derrida and Hägglund initiate.

As I will further argue throughout this study, the core significance of a technological aesthetic of imperfection lies in its intimation of imperfection, finitude, and fragility as both universal *and* materially specific conditions. Imperfection denotes a frictional aesthetic that can attune the user to the way in which autoimmunity constitutes both technology and the worlds it facilitates. What if, through an aesthetic of imperfection, technology would more effectively relay its finite composition? What kinds of chronolibidinal attachment and what forms of care could this inspire? What would the wider pharmacological significance of such investments be in relation to contemporary technological processes? The artworks I analyze in Chapters 3 through 5, through their aesthetic of imperfection, offer a possible answer to these questions. First, however, it is essential to conceptualize the pharmacological setting in which such an aesthetic contemporarily emerges, paying particular heed to the finite and fragile conditions that support today's purportedly frictionless digital technosphere. This is the aim of Chapter 2.

2

A Pharmacology of Frictionlessness

Digital Destructions and the Frictional Value of Imperfection

In her 2020 memoir *Uncanny Valley*, American writer Anna Wiener portrays her years spent working in Silicon Valley. She recounts how her initial gold-rush euphoria was gradually displaced by a disenchantment with the area's privileged culture and rampant libertarianism, which would ultimately provoke her withdrawal from the scene. At heart, the book is a document of growing unease about the technological state of the world—one passage in particular sums up the pervasive mindset that caused Wiener to turn her back on Silicon Valley:

> The endgame was the same for everyone: Growth at any cost. Scale above all. Disrupt, then dominate. At the end of the idea: A world improved by companies improved by data. A world of actionable metrics, in which developers would never stop optimizing and users would never stop looking at their screens. A world freed of decision-making, the unnecessary friction of human behavior, where everything—whittled down to the fastest, simplest, sleekest version of itself—could be optimized, prioritized, monetized and controlled. (2020: 136)

This excerpt epitomizes what I will conceptualize as the perfection-oriented design philosophy of *frictionlessness*. This philosophy heavily informs how contemporary digital technology, through its functional and aesthetic design, impacts user perception and therefore forms the dominant backdrop against which technological aesthetics of imperfection presently take shape.

In the previous chapter, I principally theorized imperfection as an existential condition that reveals finitude to define both humans and their machines. In this chapter, I heed Bernard Stiegler's warning that, under the toxic light of today's technical constellations, it is insufficient to only deconstruct metaphysical or existential presuppositions (2013: 50). What is required, rather, is to subsequently develop technological and political critiques that consider how existentially undifferentiated conditions play out materially, often to highly unequal effect (2013: 50). Earlier, I introduced the notion of pharmacological thinking—of conceiving of technology as simultaneously poison and cure, variably expanding and impeding human perceptual, sensory, and affective capacities—and stressed that one of its prime values lies in analyzing how the material conditions of technological production are aesthetically communicated to the user. Pharmacological perspectives should, in other words, grapple with the material foundations of technologically mediated experiences and with the way these foundations are eclipsed or foregrounded in user perception. This chapter offers a pharmacological reading of the philosophy of frictionlessness that focuses on its aesthetic qualities and explicates how this philosophy, which strongly determines how technology is utilized and perceived, has inaugurated a poisonous logic. Frictionlessness precipitates but at the same time seeks to aesthetically evacuate destruction from its designs, shaping a pharmacological situation in which users are granted only a limited glimpse of the toxins that seep from their devices. A technological aesthetic of imperfection, so I contend, poses a possibly curative counterweight to frictionlessness— imperfection comprises an element of friction that conveys a sense of technological finitude and fragility, potentially shaping a more care-inflected relation to technology.

The argument I develop in this chapter is structured as follows. I first define frictionlessness as the prime design philosophy of Silicon Valley and explain that it aesthetically and ideologically solidifies the operations of surveillance capitalism and platform capitalism. I then describe how this philosophy expands itself by shaping user experiences that appeal to the values of user-friendliness, connectivity, and optimization. This enables me to chart the pharmacological qualities of frictionlessness, which I do by revisiting the concept of spectrality from Chapter 1. More exactly, I identify three forms of spectrality that, in conjunction, elucidate the most toxic pharmacological ramifications of purportedly frictionless technologies: frictionlessness increasingly hides technology from the user's view (*technological spectrality*), which in turn both expands and obscures its material foundations (*spectralization*) and prevents the development of aesthetic alternatives (*hauntological aesthetics*). Following from these claims, I present a technological aesthetic of imperfection as a source of friction that, by highlighting finitude and fragility, carves out conditions for new, more pharmacologically and chronolibidinally sustainable relations to technology to emerge.

Frictionlessness and Silicon Valley: Surveillance Capitalism and Platform Capitalism

"Friction"—a term to which I will attend more closely at the end of this chapter—is defined by anthropologist Anna Lowenhaupt Tsing as "the awkward, unequal, unstable, and creative qualities of interconnection across difference" (2005: 4). Frictionlessness, conversely, describes the (seeming) absence of friction: a situation in which agents interface consonantly, difference ostensibly dissolved into a seamless whole. In this book, I approach frictionlessness as a design philosophy that characterizes the conduct of Silicon Valley and particularly of the industry giants that comprise Big Tech—Alphabet (Google), Amazon, Apple, Meta (Facebook), and Microsoft. Frictionlessness governs the design of smartphones, social media platforms, streaming services, algorithmic recommendation systems, the applications that drive the gig economy and the smart devices that today dot many a home. Intuiting this pervasive logic, media scholars Olia Lialina and Dragan Espenschied state that "the computer's ultimate purpose is to become an invisible 'appliance,' transparent interface and device denying any characteristics of its own. Most computing power is used in an attempt to make people forget about computers" (2009: 9). Indeed, the philosophy of frictionlessness maintains that the perfectibility of consumer technology lies in designing appliances that function so smoothly and that are woven so seamlessly into the fabric of everyday life that the space between a user's emergent and often technologically informed needs and their technological satiation is optimally reduced. According to this vision, even the most mundane tasks—whether it concerns ordering food, checking in with friends, or switching on the lights—should be mediated by technologies that have been designed to remove friction from experience.

This aim is aesthetically redoubled by the outward design of digital devices, which are generally made to look lustrous and sleek, stripped of imperfections and blemishes. The spotless surfaces of smartphones and laptops bolster a flawless aesthetic that suggests a sensuous smoothness of functionality as well as a sterile process of production. Moreover, the representational capacities of today's technologies yield increasingly lifelike images: new video game consoles and 8K resolution screens promise a pristine and fully immersive experience that will make the user forget about the very fact of technological mediation. Frictionlessness thus touts an aesthetic logic in which immaculate devices enable crystal-clear transmissions while increasingly receding into the distance of perception.

Frictionlessness as I understand it is primarily about technologies' functional design and aesthetic qualities and less about the specific content digital devices distribute; frictionlessness mainly concerns the transparency, intuitiveness, and efficiency of technological operations. It is, above all, a means of technologically impacting (or, as Lialina and Espenschied's words imply, not impacting) perception and a way of dissuading users from considering the digital's material underpinnings and illicit operations. As elaborated in the previous chapter through Stiegler's concept of tertiary retention—referring to the technical and exteriorized element of human consciousness—and the notion of postphenomenology, technology co-composes perception and cognition. It is therefore urgent to ask which modes of apprehending the world are atrophied when the eradication of friction from user experience is made paramount.

First, however, it is informative to contextualize the ascendancy of frictionlessness so that we better understand its aims. To be sure, the goal of having things function unobtrusively has always been central to the design of consumer technology (and, if we follow Heidegger, the capacity to disappear by becoming present-at-hand is a defining quality of technology in general). Yet, the rise of computational technology has offered companies unprecedented means of designing for inconspicuousness and efficiency, uncovering new ways of generating revenue in the process. Tellingly, computer scientist Mark Weiser, often credited as the godfather of ubiquitous computing, opened his influential 1991 article "The Computer for the 21st Century" with the following words: "The most profound technologies are those that disappear. They weave themselves into the fabric of everyday life until they are indistinguishable from it" (78). Weiser went on to paint an idyllic picture of twenty-first- century life—a life utterly mediated by computers that are hidden in miniscule tabs and badges while regulating all sorts of banal activities, from making coffee to reading the newspaper. In Weiser's view, the logic of "pushing computers into the background" portended a "new dominant mode of computer access" that, "by making everything easier and faster to do, with less strain and mental gymnastics, . . . will transform what is apparently possible. . . . Ease of use makes an enormous difference" (89). While acknowledging some of the privacy issues such a pervasive computational apparatus could introduce, Weiser remained utterly enthusiastic about this image of mundane life in which the omnipresent-yet-ethereal tendrils of computation would bring vast gains in convenience and productivity. Here, then, we find an initial vision laid out for what would eventually become Silicon Valley's dominant design philosophy of frictionlessness.

Another seminal moment for the philosophy of frictionlessness was the 1995 release of Bill Gates's *The Road Ahead*, a book in which the cofounder and former CEO of Microsoft presented an influential account of how the internet would transform the world. For Gates, the widespread availability of and rapid developments in information and communication technology would "carry us into a new world of low-friction, low-overhead capitalism, in which market information will be plentiful and transaction costs low[:]"

> Capitalism, demonstrably the greatest of the constructed economic systems, has in the past decade clearly proved its advantages over the alternative systems. The information highway will magnify those advantages. It will allow those who produce goods to see, a lot more efficiently than ever before, what buyers want, and will allow potential consumers to buy those goods more efficiently. (181–3)

Implicitly merging Weiser's computational views with a desired optimization of consumer capitalism, Gates declared that digital technology—promising full personalization, a proliferation of "electronic efficiency," transferability "at the speed of light" (180–1)—would serve to rid capitalism of its perceived market frictions, capacitating both consumers and producers to make more informed and efficient decisions.

Gates's account presaged the rise of friction(lessness) as a discursive and aesthetic preoccupation within technological design (see also Kemper and Jankowski forthcoming). From Microsoft's boastful announcement of "friction free software" (Microsoft 1999) and later introduction of the "Frictionless Devices Initiative" (Microsoft 2020) to Amazon's spearheading of a "radically more friction-free economy" (Colvin and Derousseau 2017) and from Google's "maniacal focus" on reducing "every possible friction point" (Adams 2013) to Meta's recent promise of a "Zero Friction Future" that will gratify all communicative and consumerist desires at "the speed of now" (Facebook IQ 2017): over the past decades, friction has become a dreaded specter that haunts the mind of many a tech entrepreneur. Frictionlessness, by contrast, is envisioned as a desired state in which the jagged edges of cultural practice and human–technology interaction have been smoothened. The ascendance of frictionlessness as a design principle is perhaps evinced most clearly by the entrepreneurial handbook *Frictionless* (2020), in which Christiane Lemieux and Duff McDonald tout the mindset and technological ability to reduce friction as the ultimate markers of success: "We're talking about an ontological change here, folks: we have entered a new era, and frictionlessness is the state of being that we're all aiming for, whether it's as individuals, groups of people working together, or in the institutions in which we put out collective faith" (Lemieux and McDonald 2020). These are but some of the innumerable discursive denouncements of friction, impacting anything from interface design to business plans, that suggest that frictionlessness has become a pervasive design philosophy within Silicon Valley.

Understanding frictionlessness as a design philosophy entails viewing it not as an innocent trend but as a force that actively conditions the world and the economic, ecological, and cultural relations within it (Latour 2009). In this light, the link that Gates draws between friction and capitalism is significant but does not yet fully capture the contemporary relation between frictionlessness and capitalist modes of production. Frictionlessness as it reigns today, I argue, essentially forms the aesthetic logic that emerges from and reinforces the collusion of surveillance capitalism and platform capitalism. Surveillance capitalism is, as Shoshana Zuboff (2019) describes it, a new form of capitalism that came to fruition after the dotcom crash of the late 1990s and early 2000s, in a time when tech companies desperately needed new sources of profit to stay afloat. Sensing that more could be done with the user data they gathered than simply improving their search engine, Google was the first company to discover the untapped potential of behavioral surplus (Zuboff 2019: 74–5). The company found that their vast data stores effectively enabled them "to read users' minds for the purposes of matching ads to their interests, as those interests are deduced from the collateral traces of online behavior. With Google's unique access to behavioral data, it would now be possible to know what a *particular* individual in a particular time and place was thinking, feeling, and doing" (Zuboff 2019: 91, emphasis in original). This behavioral stock proved highly lucrative, as Google inferred that they could intensify the accumulation of data and turn it toward the hugely profitable operation of targeted advertising. Quickly, the logic of surveillance capitalism started to spread through the industry, most notably prompting Facebook to construct its own data empire (Zuboff 2019: 91–2).

Discovering the value of behavioral surplus confronted these companies with a question: What are the most effective means to gather data? Zuboff regularly employs the term "friction" to explain how tech companies grapple with this conundrum. Friction, when seen from the viewpoint of Silicon Valley, concerns humanity's intransigent behavior as well as its stubborn regulatory frameworks, both of which pose considerable obstructions to the steady accumulation of data. How to create environments in which people reliably and freely turn over their data to their machines? How to realize the dream of an unimpeded, fully "frictionless flow of behavioral surplus" (Zuboff 2019: 105)? As Zuboff explains, part of the solution to this question lies in creating a world, inspired by Mark Weiser's visions, in which "the computer would be operational everywhere and detectable nowhere, always beyond the edge of individual awareness," continually collecting data while remaining indiscernible (2019: 227). Here, Zuboff broaches the simultaneous movement of extension and obfuscation that marks the core logic of frictionlessness: in order to maximize surveillance assets, digital technology must become infrastructural, in the sense that it should be indispensable to the navigation of everyday life while receding from perception and consciousness (cf. Star 1999), becoming a kind of unconscious operating system that underpins all human activity. This logic is, for instance, encapsulated by the multitudes of smart technologies—formerly mute objects like refrigerators, toothbrushes, and doorbells that have been turned into reticular extractors of data—that ideally function as an interoperative assemblage that ensures no drop of domestic data escapes the grasp of Big Tech (Zuboff 2019: 239; see also Sadowski 2020).

Surveillance capitalism colludes with what Nick Srnicek (2017) has termed "platform capitalism," an instantiation of capitalism that is equally obsessed with the logic of frictionlessness and capture of data. Platforms, in this context, should be understood in both their technological and economic definition: as computational, programmable architectures *and* multisided markets that mediate communications between different parties and stakeholders (Helmond 2015: 2). Whereas surveillance capitalism was spurred by the burst of the dotcom bubble and the subsequent rise of Web 2.0, platform capitalism was precipitated by the 2008 crash. The long aftermath of the crash saw venture capitalists forage for new investment opportunities while, at the same time, (public) services and utilities were being scaled back, allowing platform companies to step in and take over. In this precarious and flexible labor climate, people with the means to do so increasingly opted for the kind of electronic efficiency promised by platform-mediated services that offered, for instance, grocery delivery, household cleaning, and urban transportation. By reconfiguring domestic and urban practices through a logic of "X-as-a-service," platform companies ensured that everyday life increasingly took place on corporate platforms (Sadowski 2020: 61–2). Platformization (Helmond 2015), then, has become the primary way through which Big Tech renders itself infrastructural; it embodies a monopolistic tendency in which a few platforms have the power to construe vast platform ecosystems. The perennial expansion of these ecosystems grants companies unprecedented access to surveillance capital through the decentralization of data mining (seeking out new activities to mediate and services to offer) and the recentralization of data processing (feeding the collected data back into their centralized data architectures) (Helmond 2015: 5). Again, frictionlessness is key: platforms predominantly aim to "decrease market frictions" (Poell, Nieborg, and Duffy 2021: 185) by breaking the conduct of everyday life down into a series of efficient services offered through privately owned platforms, aiming for a smooth integration of ever more platform subsidiaries.

One of the ten theses that media theorist Jathan Sadowski advances on digital capitalism is that "[t]he most insidious product of Silicon Valley is not a technology but rather an ideology" (2020: 66). "Frictionlessness," I contend, is the term with which this ideology can best be conceptualized—Sadowski, not coincidentally, dubs "frictionless" the buzzword of Silicon Valley (2020: 47). From the perspective of Big Tech, frictionlessness comprises the simultaneous extension and concealment of technology, a process that is geared toward maximizing data accumulation, generating profit, and optimizing control. Yet, frictionlessness is at the same time ideologically spun as beneficial to the user. The f8 developers conference in 2011 marked perhaps the most explicit moment in the burgeoning promotion of frictionlessness as a user-oriented feature. At this event, Mark Zuckerberg unveiled Facebook's Timeline, which was framed to evince "a new aesthetic" that would usher in a new kind of user experience (Shayon 2011). The renewed Facebook, Zuckerberg professed, would grant users optimal connectivity and "real-time serendipity in a friction-less experience" (Shayon 2011). Notably, Zuckerberg's glorification of a frictionless user experience reveals how the reduction of friction that guides both surveillance capitalism and platform capitalism is sold as desirable and as something that users should actively seek out. But of what exactly is the user experience that frictionless technologies promise composed and what are its pharmacological implications? I suggest that Silicon Valley's aims are ideologically repackaged through three central design values that pharmacologically define the user experience of frictionlessness: *user-friendliness*, *connectivity*, and *optimization*.

Frictionlessness and the User: User-friendliness, Connectivity, and Optimization

These values, framed to primarily benefit the user, furtively serve to solidify and widen the reach of frictionless design and its underlying economic model. I will focus mostly on how these values are technologically entrenched to expand frictionlessness, but it is important to note that they also speak to wider cultural imperatives with a history that extends beyond the technological. Each value serves an implicit function within the paradigm of frictionlessness. User-friendliness is a term that signals how technological operations are increasingly concealed in view of the ostensible goal of improving user experience, thereby legitimizing an increased spectrality or imperceptibility of technology. Connectivity is a means of promoting seamlessness and of broadening the infrastructural influence of technology.

The primary role of optimization is to establish a drive for a technological perfection that is always just out of reach and to shore up a logic of constant consumption. Each of these three concepts has warranted multitudes of studies but here I only roughly outline them; I draw on them to explain how the characteristics of frictionless user experience are shaped and reinforced, laying the groundwork for a more extensive pharmacological analysis of frictionlessness and its poisonous implications.

The value of user-friendliness, combining convenience and comfort, places questions of immersion and intuitiveness at the heart of digital design. One of its most obvious aims is to preserve a data-driven attention economy that tethers users to their myriad devices, shaping experiences that absorb the user to such an extent that she is discouraged from directing her attention elsewhere, away from technology (Bueno 2017). Interfaces and features are to make as little aesthetic and cognitive demand as possible so that people can optimally consume the content and services technology disseminates. In the present context I am, however, more interested in a related function of user-friendliness: user-friendliness is the primary means for tech companies to justify, through an appeal to the reduction of friction, an increased sense of spectrality that characterizes both technology and its conditions of production. As will shortly be discussed in more detail, spectral associations are often projected onto those technologies whose functioning eschews vision and whose operations elide cognition. The terminology of user-friendliness allows companies to claim that these intensified processes of occultation are realized in the interest of the user. It enables designers to declare that they have not remodeled their devices to tie users more deeply to their data- and surveillance-driven profit model or to restrict user agency; rather, they have merely designed their technologies to be more user-friendly and convenient by removing possible sources of friction. What user could object to a friendlier device, and how better to reduce the risk of friction than to hide a plenitude of technological operations from the user's gaze altogether?

While I am mainly interested in user-friendliness as a design principle that fetishizes frictionless technological experiences, it should be noted that it is also paradigmatic of wider cultural processes. The notion of user-friendliness, so digital humanities scholar Alan Liu explains, registers the spread of a cultural logic that extols ease-of-use as something not merely to be expected from machines but rather as a principle that should guide all forms of communication: to live and labor in a world of user-friendly applications is to be bound, "through automatic participation in a universal environment of 'user friendliness,' to corporate culture as the stage of general culture, as the new model of general sociality, interaction, and communication" (2004:

172). Liu, in other words, suggests the codification of user-friendliness as a technological design value to be part and parcel of the codification of user-friendly societies, whose citizens are envisioned to communicate without the frictions of noise, miscommunication, and personality. This cultural rationale and its consolidation in frictionless devices can be read in line with important theorizations about the cultural fetishization of happiness and positivity over and against outward signs of misery and malaise (Ahmed 2010; Ehrenreich 2010). At the same time, and in fact often precisely for this reason, the smooth and easily navigable interfaces of frictionless technologies turn out to be wellsprings of anxiety. Think, for example, of the unhealthy perfectionism attendant on Snapchat's polished selfie-culture (Pisters 2021); of how feelings of sadness and depression have become an integrated feature of supposedly user-friendly social media platforms (Lovink 2019); or of discussions of YouTube as a gleaming engine of radicalization (Ribeiro et al. 2020). While such specific phenomena are vital to note, they are ultimately not at the forefront of my analysis; my focus is on the pharmacological implications of the aesthetic and functional logic of frictionless design itself.

The second value that is constitutive of frictionlessness is connectivity. Media theorist José van Dijck presents connectivity as the ideological principle that unites major social media platforms[1] by shaping an imaginary of the digital as a space designed for interconnection; a realm of "frictionless online traffic" (2013: 21) and "frictionless sharing" (2013: 65) where people can freely communicate and exchange.[2] The value of connectivity thus visualizes the world, in the words of influential digital industry executive Ted Cohen, as a "nirvana of interoperability" (quoted in Van Dijck 2013: 164).

---

[1] This logic of connectivity and its economic underpinnings are not simply reducible to the rise of these new platforms. They have a longer prehistory that is recounted in sociologists Luc Boltanski and Eve Chiapello's influential *The New Spirit of Capitalism* (2007)—a book curiously absent from Van Dijck's analysis. The notion of connection forms a crucial node in Boltanski and Chiapello's theorization of capitalism as an adaptive force, capable of transforming parts of its functionality to sublimate critique into new means of generating capital. Specifically, Boltanski and Chiapello map the inauguration, after the student protests of 1968, of a "connexionist world" (2007: 131)—a world in which the forging of ties, the multiplication of connections, and the navigation of networks have been elevated as guiding principles in the conduct of life.

[2] A primary reason for this expansionist disposition is the engineering and stimulation of the production of data (Van Dijck 2013: 12). Adhering to the project of surveillance and platform capitalism, the logic of connectivity implores a constant search for new sites to mine, conceiving of the world as an endless reserve where all that exists should be made to "radiate data" (Berry and Dieter 2015b: 3). While I cannot overstate the centrality of data to many of the digital processes I analyze here—data today serving as capitalism's chief engine of profit and growth (Srnicek 2017: 6)—the term "data" will only be latent in what follows. This is primarily because the notion of data is not directly relevant to the experiences afforded by my case studies and to the aesthetic of imperfection I analyze.

Today, this interoperability does not just extend to users but also concerns the interlinking of devices, applications and objects. So-called smart technologies replace the isolated devices of yore—from vacuum cleaners to refrigerators and from thermostats to toothbrushes—and turn the twenty-first- century household into an assemblage of communicative and connective products. There is an expansionist logic to this, as digital technologies, in their interconnectivity, come to mediate a growing degree of human activity, construing an "everyday life" that is "always already computational" (Berry and Fagerjord 2017: 16). Indeed, networked connectivity, and the devices needed to realize it, today occupy an infrastructural status; connective technologies are indistinguishably tied up with people's power to act and have impact, taking shape via smart devices, platform-mediated services, and social media alike (Paasonen 2021: 55). In the frame of the present chapter, the most important function of connectivity is, then, to generalize a functional and aesthetic ethos of seamless digital expansion. Frictionlessness boasts a logic of effortless connection and extension, broadening the scope of its data-gathering activities while ridding the connections it creates of perceptible friction.

The cultural drive for more and faster connections, for more efficient actualizations of agency, discloses the last component that defines the user experience of frictionlessness: optimization. This logic encapsulates the conception of perfection that is operative behind the desire for frictionless technology. In a recent influential essay titled "Always Be Optimizing," writer Jia Tolentino identifies optimization as a pervasive imperative: "Today, the principle of optimization—the process of making something, as the dictionary puts it, 'as fully perfect, functional, or effective as possible'—thrives in extremity" (2019: 82). Her account reveals the preponderance of a cultural mood that discourages people from becoming stagnant and instead urges them to seek out ways of being more productive, living more efficiently, and exuding more vitality. Optimization is, to phrase this differently, a constant and performative process in which perfection is something that is always sought but never reached. By implication, optimization is inextricably bound to consumption: it stimulates an insatiable passion for new products and services that afford fresh avenues of productivity, connectivity, and efficiency.

While Tolentino's essay takes the measure of Western society in general (analyzing popular phenomena such as barre fitness, athleisure, and self-help manuals), media theorist Wendy Hui Kyong Chun's phrase "updating to remain the same" reveals that optimization also reigns in the realm of the digital (2016). Digital technologies, so Chun describes, are driven by ephemerality: new updates and devices arrive without cease, and, in their technological dependence, users are perpetually encouraged, if not outright required, to purchase and/or install such optimized commodities. Technologies are quickly outmoded, always existing precariously on the precipice of obsolescence, but instead of prompting practices of maintenance and repair (cf. Jackson 2014), frictionlessness prescribes rejection and replacement through consumption. Frictionlessness cultivates, in other words, a vigorous but fleeting investment in its individual devices, wherein desire is always ready to be displaced onto a new, more connective or user-friendly object. This situation traps users in a "never-advancing present" (Chun 2016: 76)—a present systematically punctuated by small moments of crisis that necessitate the consumption of another new update, even though this does not build toward qualitative difference; the logic of optimization and its implicit promise of desuetude remain cardinal.[3] Things can always be more optimal; more friction can always be removed. Optimization thus stimulates a logic of digital consumerism that expedites Silicon Valley's surveillance- and revenue-based aims by inspiring a perpetual need for more devices and more data.

The twinned dynamic of optimization and consumption unveils the peculiar ontological status that perfection holds within the paradigm of frictionlessness. To design for frictionlessness is not to anticipate the ultimate possibility of a perfectly frictionless technology. Rather, it is to understand perfection as something that necessarily remains evasive. A similar argument is developed by sociologists Vera King, Benigna Gerisch, and Hartmut Rosa, who, in their recent edited volume *Lost in Perfection* (whose subtitle reads "impacts of optimisation on culture and psyche"), trace a distinctly technological picture of perfection:

> We are witnessing a shift away from the unattainable moral-cum-aesthetic ideal of perfection with its implicit recognition of the relative imperfection of everyday practical life and the limitations imposed on us as mortal human beings. The new norm is infinite optimization, pushing back the limits further and further and permanently transcending those limits as we do so. (2019: 3)

This account aligns with philosopher Franco "Bifo" Berardi's claim that perfection and imperfection today carry less the theological connotations

---

[3] Theorizations that further underline how optimization is about augmenting and intensifying what already exists rather than inviting the truly new and different have been set forth by Stiegler and Derrida, both of whom see in today's mantras of constant innovation a solidification of the status quo rather than a prospect of radical change (Derrida 2007a: 22–3; Stiegler 2019: 32).

of an irreparably fallen state and more the injunction of a something-to-be-done here in the present, in this world, with the aid of technology (2011: 25). This injunction does not, however, dispel the indelible primacy of imperfection that I theorized in the previous chapter; this existential condition remains the foremost reason that anything like a final perfection, both within the paradigm of frictionlessness and outside it, is impossible. In fact, the new material conception and experience of perfection *rely* on the primacy of imperfection to operate; the conditions of imperfection and finitude are precisely what ensure that technologies will never culminate in a state of absolute perfection and will therefore remain optimizable. These conditions are even actively mobilized by companies to guarantee the constancy of consumption: the practice of "planned obsolescence"—a term that will return in Chapters 3 and 4—reveals that finitude is frequently an integrated design feature. Optimization, in sum, facilitates a user culture of technological consumption by making the parameters of (im)perfection highly plastic and transposable.

A Pharmacology of Frictionlessness: A Philosophy of Ghosts

Having established how Silicon Valley's extractivist model is consolidated through the user experiences it designs, we can now inquire into the pharmacological effects that a user experience defined by user-friendliness, connectivity, and optimization produces. Recalling last chapter's argument that a sustainable pharmacologic approach should consider both technology's perceptual affordances and the elements that facilitate those affordances— technologies and the user experience they shape are always dependent on assemblages of actors and matters of which the user may or may not be made aware—I will focus especially on how the material conditions of frictionlessness are impacted by and aesthetically transmitted to the user. As I have described elsewhere, "[t]he approach of pharmacology . . . comprises an analysis of how a technology's toxins and tonics materialize through the forms of perception and attention the technology in question facilitates" (2022: 60). Such an approach involves a consideration of the network of actors that facilitate the experiences that *pharmaka* create and the extent to which the user is alerted to, and prompted to care for, the plight of these actors. For this reason, my pharmacological analysis will only be moderately concerned with the user's individual susceptibility to surveillance and control, central though these concerns are to frictionless technologies. Not only have these effects been well-documented elsewhere (see, for example, Zuboff 2019; Stiegler 2019; Lovink 2019; Berardi 2015), but adopting a lens that only addresses the individual user's psychopathological composition misses what I consider most pharmacologically significant about frictionlessness: that both its underlying economy and its perceptual affordances are ecologically premised on a simultaneous expansion and obfuscation of destructive material procedures. On the whole, then, my argument is that the seemingly curative augmentations and alleviations of perception that frictionlessness brings are directly linked to a toxic reduction of the user's ability to sense the digital's underlying, expanding systems of exploitation.

What I will primarily suggest in the sections that follow is that in order to develop this argument, it is instructive to conceive of frictionlessness as aesthetically and functionally breeding a thoroughly *spectral* culture. In the previous chapter, I introduced the concept of spectrality primarily as a constitutive condition that registers the disjointed nature of time. The general structure of spacing that I discussed via Jacques Derrida and Martin Hägglund imparts to each moment the possibility to haunt: should a moment inscribe itself in some way onto a material stratum, it is given, by something external to it, the chance (but never the assurance) of remaining as a trace for the future. This contaminative process, in turn, allows for the apparition of historically, culturally, and technologically specific ghosts. In line with the disjointed and pollutive nature of spacing, ghosts are figures that corrode clear boundaries, straining the membrane that divides absence from presence, life from death, visibility from invisibility. Ghosts can serve as markers of finitude, but at the same time suggest the impossibility of fully putting things to rest; in their ability to haunt from beyond their ostensible grave, ghosts complicate ideas of living and dying. If aesthetic imperfections frequently appear indexical to time's relentless passage, ghosts, too, unveil the reality of loss and the fragile composition of even the most durable of matters. In their recurrence, however, they also reveal the perseverance of the past, perhaps confronting one with debts that remain unsettled.[4] Ghosts, moreover, can arrive from the future, not

---

[4] For examples of ghosts as traces of the past that persist in the present, see Robert MacFarlane's Underland (2019), wherein he draws on the concepts of ghosts and hauntings to make intelligible how the extractions and toxic depositories that sustain the modern world are carefully kept from view while increasingly creeping to the surface; Eve Tuck and C. Ree's "A Glossary of Haunting," which invokes the ghosts of settler colonialism to argue that haunting is a condition that permanently imposes itself on the United States as an ineffaceable stain shaped by slavery and genocide (2016: 642); and, relatedly, Thomas Pynchon's ghost-ridden novel Mason & Dixon, whose pages are possessed by a host of ghosts begat by the supposed Age of Reason: "But here is a Collective Ghost of more than household Scale,—the Wrongs committed Daily against

only as portents of a prospective finitude but as projections of a world to come that already lays its claim on the present.[5] By implication, ghosts are often framed as invisible or virtual agents that nonetheless generate material effects.

This rich web of qualities and associations has made the ghost an explanatory figure in cultural and literary studies that helps to conceptualize how life and death, virtuality and actuality, past and future are materially active in a historically situated present (see, for example, Peeren 2014; Gordon 1997; Tanner 2016; Fisher 2014). The figure of the ghost, by virtue of its associations with temporality, virtuality, and finitude, can also be productive when evaluating technological processes. More precisely, there are three ways in which spectrality has been conceptualized that I suggest to be especially valuable for analyzing both the pharmacological situation frictionlessness produces and the possible counterweight that a technological aesthetic of imperfection poses. The first of these forms of spectrality concerns what I call *technological spectrality*, or the technological tendency to circumvent cognition and perception, a phenomenon that has far-reaching implications for any discussion of technological aesthetics. The second form employs the idea of *spectralization* to describe processes that render subjects both vulnerable and invisible, enabling a conception of frictionlessness as a machinery that functions by producing ghosts. The third form assesses the critical value of *hauntological aesthetics* as an aesthetic- and imperfection-based means of probing technology for alternate pharmacological futures by inquiring into the intersection of technology and temporal perception. The first two forms capture the pharmacological nature of frictionlessness whereas the latter form helps to decode the critical value of imperfection.

To illustrate the extent of these ghostly forms, let us take a brief look at one of the crowning achievements of frictionless design: Amazon's Echo.[6] Amazon Echo is a consumer brand of voice-controlled smart speakers that

---

the Slaves, petty and grave ones alike, going unrecorded, charm'd invisible to history, invisible yet possessing Mass, and Velocity, able not only to rattle Chains but to break them as well" (1998: 68).

[5] For examples of ghosts as agents of the future, see Mark Fisher's work on the effects of (lost) futures on Western culture (2014) and Patricia Ticineto Clough on capital's spectral power, defined by the speculative quest for usurpation of whatever "chance for difference comes from the future, the virtual" (2004: 19).

[6] It is worth noting that Ruhi Sarikaya (2018), the director of applied science for Alexa AI, explicitly frames Echo and its voice assistant technology Alexa in terms of frictionlessness. "Amazon", he explains, "is obsessively focused on reducing or eliminating friction—think one-click ordering, Amazon Prime or Amazon Go." Alexa is "similar to any other Amazon service" in that its primary goal is "removing friction in . . . customers' interactions with the physical and digital world" (2018). See: https://www .amazon.science/blog/making-alexa-more-friction-free.

aims to render frictionless such commonplace tasks as checking traffic, ordering groceries, and regulating room temperature. As a technology, Echo abides by the tenets of frictionlessness, displaying a sleek yet inconspicuous aesthetic while offering an array of convenient functions. At face value, it is a highly user-friendly device: it enables users to complete a wealth of domestic tasks via the minor exertion of vocalizing one's demands, hiding a plenitude of technological operations from view in the process. Moreover, Echo seamlessly links disparate objects, services, and activities into one small, concentrated hub and thus operates on a logic of seamless connectivity. Finally, Echo comes with a constant stream of optimizations and updates, sentencing numerous other technologies to obsolescence—notably, Amazon has discontinued its own line of Dash products in purview of Echo's growing capacities. In their ability to carry out a myriad of tasks with only a negligible amount of effort required, Echo products might seem to operate almost immaterially.

One glance at the map that comprises Kate Crawford and Vladan Joler's impressive "Anatomy of an AI System"[7] (2018) should swiftly dispel such notions. This map illuminates the colossal system of material extraction, exploited labor, and large-scale transportation that covertly composes Echo's ongoing conditions of possibility: the diagram depicts the indispensable activities of a variegated and often undervalued workforce (consisting of component manufacturers, smelters and refiners, miners, waste collectors, software developers, etc.); the immense amounts of energy and water that are needed for AI training; the maintenance and extension of the submarine cable infrastructures without which Echo would not be able to function as frictionlessly as it does; the physically hazardous practices of waste collection, dismantling, and disposal impelled by discarded devices; and the resultant deep time trajectories of geological disintegration and destruction. The sanguine voice of Echo's virtual assistant Alexa does, however, not speak of these tortuous origins, nor of the future calamities to which it might contribute. Crawford and Joler's project is effectively an exercise in bringing to light how Echo's "shiny design options" seek to conceal that "each small moment of convenience [or frictionlessness]—be it answering a question, turning on a light, or playing a song—requires a vast planetary network, fueled by the extraction of non-renewable materials, labor, and data" (2018). Echo, in obfuscating its activities and material foundations, typifies the spectral logic that marks the philosophy of frictionlessness: frictionlessness must aesthetically conceal a legion of ghosts in order to function as smoothly

---

[7] This map can be found at the following link: anatomyof.ai/img/ai-anatomy-map.pdf.

as it does, giving way to a pharmacological condition that gathers three particular forms of spectrality.

A. Technological Spectrality

To understand the pharmacological situation frictionlessness produces, it is first necessary to theorize, through the figure of the ghost, how technology increasingly hides itself from view. Metaphors of the spectral and the supernatural have long followed developments in the field of science and technology (Blanco and Peeren 2013: 200). More specifically, ghosts, as imperceptible entities that nonetheless produce material effects, have frequently been invoked to indicate technological processes that escape perception and cognition. The trope of the "ghost in the machine" used to indicate computational glitches is a recent example, one that will be further explored in Chapters 3 and 4, but the cultural tendency to associate technology with preternatural power goes back much further. Jeffrey Sconce traces this history in his book *Haunted Media* (2000), covering a great miscellany of technological innovations that in one way or another came to be heeded as spectral and uncanny, as being governed by an enigmatic, sometimes even sinister, force. Indeed, is it not one of the well-worn cinematic conventions of the horror genre that a defective technology (a flickering light bulb, a static-plagued tv) announces the presence of a ghost?[8]

These mental connections between the ghostly and the technological remain pertinent today, in a time when technological capacities keep expanding. As María del Pilar Blanco and Esther Peeren explicate in relation to spectrality and technology, "[t]he proliferation of new machines in the modern age reminds us of the body's limitations (and the need to overcome them), as it can also remind us of the borders of life itself" (2013: 200). The spectral qualities that the digital machines saturating today's market may harbor confront users with areas where their power, perception, or understanding falters. While none of this should give recourse to alarmist visions of human obsolescence, the notion of *technological spectrality* pinpoints particular zones of tension within the fold of human–technology interaction. The dimension of spectrality is amplified precisely at those points where people feel their perceptual or cognitive capacities outstripped by technology; technology tends to invite spectral associations when it acts independently of the user's command, when its functioning eschews

---

[8] Literary scholar Marc Olivier even identifies a cinematic genre of ghost stories that he dubs *glitch gothic*. Within this genre, technological malfunction is fundamental to the narrative arc of the film and hints at the presence of vengeful spirits (2015: 253).

perception, or when it operates in timeframes that are irreconcilable with human time-consciousness (of course, these three experiences regularly turn out to be interrelated).[9] To be sure, technological spectrality never exists beyond the material; even the fastest, most transparent technologies operate according to an "extremely reduced 'différance'" and are reliant on spatiality in their operations (Derrida and Stiegler 2002: 129). Technological spectrality is more about what can and cannot be perceived and aesthetically experienced by human consciousness; it is, in this regard, a pharmacological concept at heart, insofar as it maps the perceptual affordances technology creates. Pharmacologically speaking, today's frictionless technologies are defined by a precipitation of ghostly encounters: "Modern technology," as one of Derrida's more well-known phrases holds, "increases tenfold the power of ghosts. The future belongs to ghosts" (Derrida and Stiegler 2002: 115).

The three central user values of frictionlessness—user-friendliness, connectivity, and optimization—propagate an (anti-)aesthetic logic of technological spectrality, whereby technology is made to function increasingly transparently, autonomously, and unobtrusively. The unrelenting technological quest for "higher resolutions, better color palette [and] screen refresh rate" (Contreras-Koterbay and Mirocha 2016: 41) is one example of the drive for technology to function ever more spectrally. As representational technologies bring more lifelike mediations, technology renders its own presence increasingly elusive, steadily receding to the peripheries of perception. With screen resolutions and pixel densities that are no longer even fully apprehensible by the human eye, the gap between the world and its technological representation appears to decrease to a point where only ghostly lines of demarcation remain.

There is, however, an even more far-reaching sense in which technology is growing more spectral in its designs—a spectrality that is epitomized by Amazon Echo, as an unobtrusive material object that carries a realm of technological activities. Intuiting the ghostly logic behind contemporary digital technologies, Wendy Hui Kyong Chun maintains that, "[a]s our machines disappear, getting flatter and flatter, the density and opacity of their computation increases," implying that the putative clarity the digital offers— perhaps at face value suggesting an eradication of ghosts—is accomplished by shielding its actual, increasingly complex operations from view (2011: 17). "The interface," Chun explains, is "haunted by processes hidden by our seemingly transparent GUIs" (2011: 60). While frictionless design may

---

[9] For a work that explores the inseparability of spectrality, technology, and perception, see Steve Goodman, Toby Heys, and Eleni Ikoniadou's edited volume *Unsound: Undead* (2019).

appear to render everyday life more orderly and manageable, it can achieve this only by broadening spectral pockets—by making its operations ever more complex while aesthetically hiding these from the user, whose interactions with a device should ideally be reduced to a modest offering of simple and user-friendly commands.

This spectral quality of today's technosphere is illustrated by a strand of media theory that scrutinizes the burgeoning capacity of digital technology to eclipse human perception and cognition (Volmar and Stine 2021). Think, for example, of the recent work of N. Katherine Hayles, wherein she theorizes cognition as something inherent to both digital systems and biological organisms, the former harboring modalities of cognition that transcend the latter's discernment (2017: 9). Likewise, Luciana Parisi conceptualizes algorithmic computation as an actual mode of thought "that cannot be reproduced or instantiated by the neuroarchitecture of the brain . . . or . . . the neurophenomenology of the mind" (2013: 170). As another example, Yuk Hui's concept of algorithmic catastrophe denotes a fractious element of contingency within the algorithmic that can trigger accidents beyond the human's knowledge or control (2015: 139). Each of these theorists presents digital technology as a spectral force that pharmacologically complicates human agency and that yields material effects even if it remains insensible. Perhaps the most cogent exposition of this intensifying technological spectrality and its relevance to my own arguments is found in media theorist Mark B. N. Hansen's notion of *twenty-first-century media*, which covers the operational logic of a host of frictionless technologies that are driven by algorithms and artificial intelligence. Hansen gives the following description of this concept:

> By twenty-first-century media, I mean to designate less a set of objects or processes than a tendency: the tendency for media to operate at microtemporal scales without any necessary—let alone any direct—connection to human sense perception and conscious awareness. . . . For the first time in history, media now typically affect the sensible confound independently of and prior to any delimited impact they many [sic] come to have on human cognitive and perceptual experience. (2015: 37)

The pharmacological tendency that Hansen outlines precludes direct aesthetic experiences of technological processes and thereby announces a profound logic of spectrality. In our impression of twenty-first-century media, time appears truly out of joint—past operations can perhaps be accessed, adjustments can be made for the future, but, because twenty-first-century media move in timeframes alien to human consciousness, a direct aesthetic relation between our own consciousness and the real-time activity of these media is impossible. The medial forms that Hansen identifies thus sit at the far end of a vast spectrum of technological spectralities; What, after all, could be more ghostly and frictionless than a technology that entirely evades perception?

In the previous chapter, I drew on Bernard Stiegler's work to suggest that human consciousness and its ability to intuit the temporal, with its attendant conditions of finitude and fragility, are partially composed through the technical mediation that comprises tertiary retention. What does it mean, then, that so many technologies are now designed to make no aesthetic demand or even to leave consciousness out of their loop altogether? While technological spectrality is in itself not a negative phenomenon—the notion of pharmacology stresses, after all, that technology always compensates for a human lack and that no technological apparatus can be glossed as unambiguously detrimental—I believe there is a particularly toxic ramification of the aesthetic situation that today's spectral technologies (do not) induce: the contemporary increase in technological spectrality is based on a deepened exploitation *and* aesthetic obscuration of the finite material infrastructures on which frictionlessness relies. In order to function ever more spectrally, technology relies on a wasteful and exploited network of resources, energy, and labor. Yet, *because* of that very same technological spectrality, these exploitations are at the same time increasingly evacuated from the aesthetic design of the machine and the experience of the user. This process defines the second form of spectrality that is central to the pharmacology of frictionlessness: *spectralization*, or the rendering invisible of fragile bodies and matters.

B. Spectralizations

In the previous chapter, I insisted that pharmacological thinking, if it wants to be truly sustainable, must assess not just how the user's individual well-being is modulated by technology (by examining, for instance, the illicit subjection to surveillance that attends a value like connectivity) but also how a technology affects and attunes its users to the other lives and finitudes that are at stake in its production. In functioning increasingly transparently, what else do today's technologies hide from the user's view? The notion of spectralization adds a vital perspective to any pharmacological reading of frictionlessness, as it allows one to render intelligible how contemporary technology operates on a twinned process of concealing *and* debilitating the lives and matters that sustain it.

Cultural theorist Esther Peeren argues that one of spectrality's most important "conceptual function[s]" is "to call attention to and assign responsibility for social practices of marginalization and erasure, and for cultural and historical blind spots" (2014: 13). This interpretation of spectrality unveils what Peeren sees as a limit in Derrida's conception of the ghost. Derrida, she explains, embodies a tendency of looking only *at* the ghost instead of through its eyes; "[t]he way the specter is," in such approaches, "predominantly focalized from the perspective of a haunted self means that only relatively powerful ghosts capable of haunting are recognized, while the truly dispossessed—those overlooked because they are considered expendable and irrelevant—remain invisible" (2014: 29). Contrary to the familiar image of an ill-starred subject beset by an indomitable ghost, many ghosts are ghostly because they are seen—if they are seen at all—as superfluous or, already in life, adjacent to death. This is illustrated by Peeren's discussion of the lives of undocumented migrant laborers as depicted in the British films *Dirty Pretty Things* (2002) and *Ghosts* (2006). While the films frame these workers as physically present, they are at the same time ignored, excluded, and even reviled, starkly exposed to the contingencies of violence. In this regard, the films' laborers are spectralized: forced to live in a ghostly state, residing in the world unheeded, existing in an everyday relation to death (Peeren 2014: 37–8).

What the existence of such living ghosts, with their all-too-real counterparts outside the filmic plots, signals is that, while the previously discussed logics of autoimmunity, imperfection, and spectrality spell a collective condition of finitude, material manifestations of finitude and fragility are multiplex and unequally distributed. Finitude and fragility, that is to say, are not lived the same by everyone. Philosopher Achille Mbembe's notion of necropolitics, denoting the general but also always locally and politically specific "instrumentalization of human existence and the material destruction of human [and, one might add, other-than-human] bodies and populations" (2019: 68; see also Peeren 2014: 51), makes painfully clear how unevenly dispersed death is. The mere exposition of collectively borne conditions is thus not enough to give way to more ethical perspectives or to a mitigation of suffering; such expositions should, rather, integrate material conditions of living and dying into their scope. One of the most pressing aspects of the necropolitical instrumentalizations that Mbembe identifies is that they are regularly exacted to support the modes of living of others. As anthropologist Didier Fassin contends, in a rectification of Michel Foucault's concept of biopower, "'to make live'—which is how biopower is usually understood—is also 'to reject into death,'" which foregrounds how certain privileged lives may be advanced at the expense of lives deemed less legitimate (2009: 54). The notion of spectrality allows one to appreciate that between the binary distinction of being made to live and being rejected into death exists a wide spectrum in which bodies and objects are, in various settings and to various degrees, impelled to exist in ghostlike liminality. Spectrality, moreover, helps to make palpable how the powers of making live and letting die are now increasingly wielded by tech companies.

Following Walter Benjamin's powerful claim that "[t]here is no document of civilization which is not at the same time a document of barbarism" (2015: 248), the design philosophy of frictionlessness, for all its seeming spotlessness and immateriality, forges its devices in furnaces of exploitation and erasure. Cultural critic Jonathan Beller underlines this when positing that today's digital technologies, covertly rooted "in the plantation, the factory, the colony, the patriarchal household, the university and the jail, reproduce and exacerbate inequality, oftentimes under the guise of a value-neutrality that tends to render their exploitative operations unconscious even if many of the resultant effects do not remain in the unthought, or the unfelt" (2018: 4). By appealing to a struggle between the sensible and the insensible, Beller's words capture the tension of a design philosophy that can tout the absence of friction only by producing and shielding a ruinous expanse from view. Frictionlessness, by extolling an immaculate aesthetic and a parallel intensification of technological spectrality, aims to remove all traces of the material violence it requires from its designs. Hence, while philosopher François J. Bonnet has recently hypothesized that contemporary digital technology prevents people from appreciating their own finitude (2020), I think it would be more apt to say that, as a feature of frictionlessness, it prevents people from recognizing the many *other* finitudes that are at stake in its production.

In this section, I propose that understanding frictionlessness as a machinery that must covertly produce ghosts in order to function enables us to better grasp the pharmacological poison of its functional and aesthetic inclinations. As indicated, I am indebted to Peeren in conceiving of spectralization as a forced imposition of spectral traits, a being made to exist as a ghost, inhabiting the present in a state of invisibility and fragility (2014: 14–15). In the case of frictionlessness, such processes of spectralization principally affect the bodies, matters, and ecospheres on whose toil the frictionless experiences of others depend.[10] By implication, I consider spectralization as

---

[10] As Peeren also emphasizes, such designations raise multiple ethical questions, as imposing a label onto a group of people to generalize their lived experiences may be seen as undesirable and objectionable, especially if this is done without their direct knowledge and consent (2014: 5–9, 73–4). I want to acknowledge this issue here and recognize that the terminology of the ghost can certainly be contested, rejected, or worked with in a different way. I believe, however, that the notion of rendering ghostly—of keeping from perception, of marginalizing and enfeebling—is the most apt and productive description to make intelligible what frictionlessness does to so many of its facilitators.

something that affects not only organic beings but material objects as well; as further expounded in Chapter 4, the notion of spectralization helps to make intelligible how (technological) matter can be rendered invisible while producing toxic effects elsewhere. The sense of spectralization this section covers is, to reiterate, directly linked to the form of spectrality I examined in the previous section; an increase in technological spectrality, making technology more powerful, frictionless, and transparent, also entails a further spectralization of material conditions, a further exploitation and obfuscation of finite resources. Yet, what, concretely, are some of the ghosts that frictionlessness hides from the user's gaze?

Despite what the ethereal rhetoric behind frictionless technology would have us believe—think, for instance, of innocuous terms like "streaming" and "cloud computing"—the roots of digital technology run deep into the earth. Research on media infrastructures stresses that the virtualizing capacities of the digital are dependent on material assemblages that constellate batteries, cable networks, radio towers, hardware accessories, and data centers, among many other material elements (Parks and Starosielski 2015; Cubitt 2017a; Riofrancos 2019). Moreover, if Crawford and Joler's map of Amazon Echo already conveyed that an apparently modest digital device like Amazon Echo runs on the ghostly fuel of mineral extraction, Crawford's more recent work even more pressingly charts the planetary costs of large-scale computation, demonstrating that digital culture is thoroughly premised on the "radical depletion of non-renewable resources" (2021: 31). In addition to relying on minerals and metals, the digital demands colossal expenditures of energy to make its operations appear weightless—there are, for instance, many reasons to assume that, in the seemingly immaterial age of streaming, the environmental costs of music consumption are at an all-time high (Devine 2019: 158–60). Likewise, today's pervasive trade in cryptocurrencies—especially those that run on the Proof of Work protocol—is entirely geared on the escalating consumption of computational power (Prasad 2021: 140). The point here is that these foundations tend to be carefully shielded from the user's gaze—lithium is covertly mined in Chile (Arboleda 2020), mass data centers invisibly support the ostensibly virtual activities of Hansen's twenty-first-century media (Hogan 2015), and our quick-fire social media interactions are carried by an undersea network of cables whose pollutive workings are absolved through the fantasy of "friction-free . . . global communication" (Starosielski 2015: 140)—even though they assert themselves across the globe in landscapes haunted by erosion, rising temperatures and, dwindling biodiversities (Tsing et al. 2017; Jucan, Parikka, and Schneider 2019). Frictionlessness, moreover, has shaped a form of perception that has so internalized the demand for speed, seamless connectivity, and optimization that when a technology *does* announce its material presence—when it lags, malfunctions, or otherwise communicates its physical nature—there is a cultural tendency to discard and replace the offending device, generalizing a logic of digital consumption and disposal that comes with rapacious material demands.

Frictionless technology not only carries considerable environmental impact; it is also structurally dependent on the exploitation and obfuscation of labor. This dependency is again brought to the fore by Crawford and Joler's map of Amazon Echo. The map makes plain that Echo, in order to deliver optimized experiences, requires both human toil *and* the removal of that toil's traces from its design. There is, in Echo's aesthetic and functionality, no tangible sign of the legion of miners, smelters, manufacturers, and AI trainers that help to create and maintain it. Frictionless design is rife with such pockets of spectralized labor.[11] One may think, for example, of the unacknowledged migrant worker systematically exposed to toxic chemicals as she cleans the semi-conductors needed to quell Silicon Valley's thirst for connectivity and high-resolution transmission (Dyer-Witheford 2015: 68–9). Or consider the precarious workforce of independent contractors that fuels the luxurious frictionlessness offered by ride-hailing and food delivery apps (Van Doorn 2017). Another salient example concerns the unsung "ghost workers" whose piecemeal and on-demand manual tasks are needed to help companies like Amazon and Google sell the image of smart, autonomous, and efficient technology (Gray and Suri 2019; see also Jones 2021).[12] These different forms of hidden labor reveal that "frictionlessness" functions as a sanitized term for a design logic that forces some individuals to work invisibly for the frictionless experience of others. Frictionless design thus operates on a cluster of judgments about which bodies, matters, and ecospheres can be marginalized and exploited if it satisfies the call for user-friendliness, connectivity, and optimization elsewhere.

---

[11] The notion of spectralized labor also gestures at a wider undervaluation of technological practices of maintenance and repair (cf. Jackson 2014; Mattern 2018). These practices, foundational to society though they are, tend to be overshadowed by narratives of technological optimization, production, and innovation. This phenomenon is something I will attend to more thoroughly in Chapter 3.

[12] While I conceive of frictionlessness mostly in terms of the transparency of technical processes and their aesthetic transmission, it is worth noting that a logic of spectralization also applies to the human labor needed to mediate the *content* of frictionless technologies. Especially noteworthy is the dusky field of content moderation work, whose laborers must sift through potentially traumatizing images in order to shield users from content that might disturb them. These workers operate, as media scholar Sarah T. Roberts describes, according to a logic of "invisibility by design" (2019: 3), as any acknowledgment of their existence would complicate the aura of frictionlessness tech companies aim to transmit.

Who are the envisioned subjects of frictionlessness and who are primarily saddled with the arduous task of facilitating the frictionless experiences of others? Who are the prime consumers of frictionless devices and who are forced to bear the brunt of the rapid turnover of technologies by residing in the toxic hinterlands of discarded devices? Frictionlessness betrays a sharp divide between the Global North and the Global South, as frictionless user experiences in the former area tend to be directly supported by destruction and exploitation in the latter. The toxicities that seep from today's modes of digital production and consumption are, as media theorist Sean Cubitt's work makes achingly plain, often felt most directly by those who inhabit occulted sites of extraction and disposal—frequently, this concerns indigenous people who, if not displaced entirely, are forced to work under inhumane conditions and to live on contaminated grounds (2017a: 40–54, 64–78). I have further charted these tensions elsewhere (Kemper 2022), but the preceding paragraphs should already suffice to disclose that, in line with what feminist data scientists Catherine D'Ignazio and Lauren F. Klein argue, the current technosphere remains driven by "capitalist and colonial forces that encourage the exploitation of Black and brown bodies so that white bodies can thrive" (2020: 184). Additionally, the field of the environmental humanities teaches us that our technological ways of living have become ways of dying for many different species (Van Dooren 2014; Haraway 2016), expanding the already vast stock of "nature's ghosts" (Barrow 2009).

These are but some of the many specters born of frictionless design and to draw up a full litany of them, or to attend to all of them in their specificity, would greatly exceed the scope of this study. The covered examples should be enough to show that the extractivist business model of frictionlessness—premised on expropriating data and constant digital expansion—is also literally extractive, perpetually demanding more minerals, labor, energy, and waste. What I am interested in here is how the logic of perceptual elision and functional indispensability that binds these various ghostly realms defines the pharmacological qualities of frictionlessness. No pharmacological project can be complete if it does not acknowledge how the individual's use and perception of technology— along with all curative effects these may be thought to bring—are today facilitated by and at the same time aesthetically emptied of a sphere of toxic conditions. Following this claim, what is pharmacologically distinct about frictionlessness is that the perceptually and cognitively alleviative terms through which its technologies are framed and experienced—connectivity, optimization, and user-friendliness, but also convenience, sustainability, immediacy, and so forth—are propped up by an obfuscated yet all-too-material realm that operates anything but frictionlessly. To counter these hidden poisons, there is a need to develop critiques that work across many different levels—from the political to the cultural, from the economic to the aesthetic.

My focus in what follows will be on the latter dimension, especially on the possibly curative effects a technological aesthetic of imperfection can bring. Considering that alternate aesthetic figurations of technology can stimulate different modes of engagements, such an aesthetic may resist the perception of technology that frictionlessness shapes, although it also has its inevitable shortcomings and should be supplemented with other critical perspectives. The prime critical value of an imperfection-oriented aesthetic is that it can challenge the way frictionlessness, through the values of user-friendliness, connectivity, and optimization, aesthetically perpetuates itself. In its expansionist inclinations and asymptotic move toward perfection, frictionlessness induces a fleeting and consumption-driven relation to the technological that belies any awareness that matter and media are not infinitely expandable, exploitable, and replaceable. Covertly, this perception is complicit in further aggravating the preceding spectralizations and spells a careless relation to technology— not in the sense that people cannot invest their devices with desires, but in the sense that these devices are heedlessly engaged and constantly overtaken and optimized, requiring the exploitation and obfuscation of more labor, more resources, more energy, more landfills. An aesthetic of imperfection can, conversely, evoke a perception of, and attachment to, technology that is more attuned to its material, finite, and contingent aspects, possibly breeding less vulnerable and noxious ghosts. The objects I analyze in Chapters 3 to 5 all draw on an aesthetic of imperfection that negates the aesthetic logic of frictionlessness and establishes a more sustained investment in technology.

While I am convinced of their significance, I cannot but admit that these objects also fall short in the face of what has been discussed. This section has outlined widespread processes of spectralization that cannot be nullified by the humble offerings these objects extend. The objects I discuss, in mostly lacking an explicitly socio-political and activist dimension, do not engage these matters in any direct sense. To those that are most harmed by the manifestations of frictionlessness, these art objects might seem unbearably trivial. Nothing here provides them any practical means of improving their situation or of attuning people directly to their plight. The value of these objects lies, rather, in challenging the vaporous agency of frictionlessness, its circular temporality of obsolescence and renewal, and its advocation of unbridled consumerism. While they do not directly address many of the specters that frictionlessness has created, these objects challenge the aesthetic logic through which such ghosts are born. More precisely, by accentuating the alluring and subversive qualities of an aesthetic of imperfection, they prompt a more invested commitment to technology. As Chapter 4 explicates, such a commitment might even entail viewing the realm of technology as itself a site of fragile ghosts that must be cared for.

To be sure, by conceiving of technology as something that can be cared for, I by no means intend to equate the objects I analyze with the hardships of spectralized laborers or the environmental erosions carried out under the banner of frictionlessness. My argument is not meant as a speculative exercise in object-oriented thought that would attribute an equal political standing to technologies as to the people and ecospheres they affect, or that would view care for a technological object as a substitute to care for living bodies. Rather, what I want to convey by conceiving of technology as something that can attract sustained emotional investment is that different ways of relating to and caring for technology can also affect technology's infrastructural composition; as feminist science and technology scholar María Puig de la Bellacasa argues, "we must take care of things in order to remain responsible for their becomings" (2017: 90). This responsibility urges us to question how sustainable forms of care can be aesthetically elicited. What aesthetic effects might counterbalance the constant production, consumption, and disposal that frictionlessness decrees? What might an attitude that tends to rather than rejects the material fallibility of technology bring, also for the many forms of spectralization that frictionlessness incurs? Could aesthetic expressions spur a pharmacological desire for entirely different technological organizations of finitude and fragility? To fully explicate the stakes of an imperfection-oriented aesthetic that shapes more curative involvements with technology, I now turn to the third form of spectrality that is pertinent to frictionlessness: hauntological aesthetics, a concept that directly underlines the critical capacity of an aesthetic of imperfection in relation to technology and spectrality.

C. Hauntological Aesthetics

Derrida's concept of hauntology has, in the past two decades, been popularized and reimagined in the context of music, art, and technology, for example through the work of cultural critic Mark Fisher (2014), music journalist Simon Reynolds (2011), author/artist Laura Grace Ford (2019), and musicologist Adam Harper (2009). These authors variably reframe hauntology as a genre, a specific sound, an effect, and/or a technological praxis. It is not my intention to settle the debate as to what hauntology actually *is*—considering the ambiguity of the ghost, it is fitting that hauntology cannot simply be defined as a delineated genre or category. I propose to think of hauntology principally as an aesthetic and pharmacological provocation that shows technology to harbor many different possible futures, urging us to probe the intimate and always malleable links between technology, temporality, and human consciousness. What I primarily want to draw attention to is, first, that artistic works described as hauntological are generally defined by a technological aesthetic of imperfection and, second, that the artistic power of hauntology should be understood in pharmacological terms by linking the work of Hägglund and Stiegler. Even though, or precisely because, the philosophy of frictionlessness poses an immediate challenge to the very possibility of hauntological aesthetics, I believe that it is crucial to underline its persisting critical capacities.

Supplementing Derrida, Fisher invites his readers to think of hauntology as "the agency of the virtual," of "that which acts without (physically) existing," and he recognizes a drive to explore such ghostly virtualities in the work of a number of musicians (2014: 18). Fisher, however, too readily equates the spectral with a notion of "pure virtuality" (2014: 22). Following Derrida's logic of the trace, there can be no agency of the virtual without a spatial component: that which acts can only act through gestures of spatialization, meaning that even a ghost cannot exist entirely unmoored from material dimensions and requires a spatial carrier to communicate. This, in fact, accounts for the profound attachment to the materiality of technology shared by the artists that Fisher enumerates. Hauntologically inclined artists probe the aesthetic, material qualities of technology to find the virtual temporalities and potentialities that lie dormant within. These artists include British producer Burial, the Scottish electronic duo Boards of Canada, and the earlier-discussed American composer William Basinski, as well as the roster of the British label Ghost Box. More recent examples are, arguably, found in the work of Demdike Stare, Tim Hecker, Amnesia Scanner, and in the genre of vaporwave that Chapter 5 discusses.

Fisher explains how "the artists that came to be labelled hauntological were suffused with an overwhelming melancholy; and they were preoccupied with the way in which technology materialized memory — hence a fascination with television, vinyl records, audiotape, and with the sounds of these technologies breaking down" (2014: 21). While he never explicitly uses the term, it is evident that one of the most prominent aesthetic traits that defines the work of these artists is the creative use of representational *imperfections*, such as vinyl crackle, static, deformed samples, and glitches. The suggestion that imperfection plays an important aesthetic role within these works is further supported by Adam Harper's observation that the hauntological aura of an artwork tends to cohere with its exhibition of a "lo-fi" (low fidelity) aesthetic, characterized by signs of "fading, dirt, or low quality materials in plastic art; noise, reverb, filters and audibly decaying or broken technology in music . . . and various forms of 'unprofessionality,' surrealism, fragmentation and collage" (2009).

The preoccupation with failing, fractured, or obsolete technology does not entail that a technological aesthetic of imperfection, loss, and decay automatically warrants a hauntological reading. What is crucial to the artworks these theorists analyze is, first, that they enact a time out of joint and, second, that this disjointedness goes beyond a mere nostalgia for the outmoded by also accommodating a futural dimension. Fisher's discussion of the elusive producer Burial, whom he views as paradigmatic of hauntology, is illustrative here. Following Fisher's reading, Burial's melancholic music evokes nothing so much as the atmosphere of a nocturnal drift through an entropic London, where vacant lots and subway stations are trembling with spectral shapes. These specters are not simply vestiges of the past but also phantasms of lost futures, symbolizing a general sensation of what Fisher, via Franco "Bifo" Berardi, describes as the "slow cancellation of the future," or the gradual disintegration of a feeling for the future as that from which the truly new can emerge (2014: 6). In the manifestation of hauntology that Burial expresses, "[w]hat should haunt us is not the *no longer* of actually existing social democracy, but the *not yet* of the futures that popular modernism trained us to expect, but which never materialised" (Fisher 2014: 27, emphasis in original). More concretely, Burial's specters are residues of a rave culture whose future-directedness was eviscerated by the digitally aggravated reality of recession and precarity. This mood is evinced through a notable aesthetic of imperfection. By disfiguring his source material (usually consisting of upbeat r&b or dance songs and video game music) and coating it with crackle and static, Burial conducts a choir of "voices under erasure" (Fisher 2014: 99), unearthing the corroded shimmers of what once conveyed hope for the future. His music, however, also suggests these futures to retain a spectral form of being that remains legible within the sonic decay, still pulsating with the faint promise of "what could still happen" (Fisher 2014: 98). Burial's music echoes sociologist Avery Gordon's suggestion that a ghost tends to simultaneously represent "a loss" and "a future possibility, a hope for the future" (1997: 64). Rather than inviting a dejected nostalgia, Burial, by way of a hauntological aesthetic, affirms the technologically affectable ability to imagine and desire alternate temporal trajectories.

Burial's aesthetic predilection for lost or faltering technologies and pop songs recalls Martin Hägglund's concept of chronolibido. In the previous chapter, I described how this concept charts the co-implication of finitude and desire; humans, chronolibido elucidates, live in a double bind to the temporal, as the capacity to care can stem only from an intuition of the possibility of loss (Hägglund 2012: 9–10). This notion also yielded a discussion of chronolibidinal aesthetics, indicating aesthetic expressions that derive their appeal from the human capacity to understand and respond to conditions of loss and mortality. Imperfection, as an aesthetic category that frequently intensifies or dramatizes the effects of the temporal, proves especially conducive to the flowing of chronolibido. Accordingly, hauntology, considering its pronounced aesthetic of imperfection, arguably embodies precisely such a chronolibidinal aesthetic. Burial's heavy use of vinyl crackle is typical; crackle draws attention to processes of mediation and at the same time connotes "a certain sense of loss" that bespeaks a "disappeared regime of materiality," creating the ghostly impression of listening to something that has long come to pass (Fisher 2014: 144). Yet, if such a focus on the technologically outmoded could easily slip into a nostalgic register, Fisher's readings reveal that hauntology tends to elicit a more temporally ambiguous kind of pathos in its listeners. Burial, in repurposing the listener's relation to technologically produced objects like pop songs and to the familiar media relaying such objects, reveals the malleable, intimate, and emotionally stimulating connection between the human mind and its technical environment. More specifically, hauntology fosters an attachment to technology as *itself* an object of chronolibido—as something that, through an intimation of temporality, loss, and finitude, may kindle an investment in its fragile and contingent state.

This quickening of the links between temporal consciousness and technology also betrays the centrality of Stiegler to hauntology's stakes, even though his name is entirely absent from scholarly work on hauntology. Stiegler, first of all, identifies a certain logic of spectrality within the general human engagement of technology; technology is what allows one to access a past one has not oneself lived through the externalized memory of past generations and thus grants a way of communing with the dead (1998: 140). More importantly, however, the hauntological artists that Fisher describes chart precisely the convergence of technology and temporal perception that Stiegler's postphenomenological perspective shows to be inextricable. Stiegler's concept of tertiary retention—the necessary, exteriorized, and technical element of time-consciousness—denotes that the technological materialization of memory and anticipation these hauntological artists explore represents a fundamental component of the structure of human consciousness (2009b: 18). The way technology materializes the temporal shapes the conditions of our perceptual and aesthetic experience. As such, recasting the aesthetic qualities of the technologies that shape tertiary retention may also engender different perceptual effects and temporal conceptions. Hauntology thus links chronolibido to pharmacology by wresting aesthetic resonance from the fragile and plastic materiality of technology and its contingent impact on temporal consciousness. By extension, hauntology shows that—even if it seems as if, following Fisher's laments, the future as a source of difference is gradually being effaced—technological trajectories are never set in stone.

If the most radical quality of hauntological aesthetics is to challenge fixed temporal trajectories by underlining technology's pharmacological and emotionally vibrant nature, what is its critical value in times of frictionlessness? While the previous two sections described forms of spectrality that frictionlessness intensifies, the notion of hauntological aesthetics encompasses a form of spectrality that is thwarted today. The challenges that frictionlessness poses to any countervailing aesthetic strategy are numerous. The immersive experiences that its devices instantiate, with their aesthetic logic of user-friendliness, connectivity, and optimization, discourage an active reimagination of technological trajectories; as Zuboff stresses, Silicon Valley's ultimate dream is to have its technologies become ubiquitously present while being "detectable nowhere" (2019: 227). Frictionlessness, by fetishizing these aesthetic proclivities, deters sustained attention for the fickle materiality of technology as it attempts to lock users into a cyclical current of novelty and obsolescence that is driven by the ever-fleeting promise of more frictionless technology. Furthermore, the notion of technological spectrality underlines that the material carriers of frictionlessness have long been shrinking in relation to the technological operations they facilitate; a growing degree of technological processes takes place entirely beyond the scaffolding of human perception and cognition. This concerted effort to minimize friction explains how, as Stiegler repeatedly insists, current generations are born into a technical milieu that works to phenomenologically restrict them from conceiving of futures beyond the addictive immediacy of a materially destructive, consumption-driven relation to the digital (2014: 22). The preceding sections have shown that frictionlessness indeed molds a form of perception that both expands and fails to aesthetically transmit technology's exploitative material conditions, thereby discouraging any emergent desire for alternatives. In light of this tendency, the most crucial capacity of a technological, hauntologically informed aesthetic of imperfection is its ability to cultivate a sense of *friction*—a sense of friction, more precisely, that emotionally attunes the user to the finitude and fragility of technology.

Friction, Finitude, Fragility

The notion of friction[13] stands in contrast to the immersive and transparent designs of frictionlessness. Certainly, an experience of friction may always lead

---

[13] "Friction," defined as "the resistance encountered when one body is moved in contact with another" (Princeton University 2010), is a term that originates in the field of physics and tribology.

the user to yield to the cycles of disposal and consumption that frictionlessness dictates—imperfection, as previously noted, is antagonistic but at the same time integral to frictionlessness, as frictionlessness perpetuates itself by rendering perfection desirable yet always elusive. Friction can, however, also trigger the user into a different mode of engagement, a mode that perhaps shows more vigilance toward technology's spectral(izing) procedures. I thus conceive of friction not as a negative phenomenon, but rather as a generative force (which is not the same as saying that all friction brings positive effects). While "friction" principally serves as an antonym of frictionless(ness), my usage of the term has also been inspired by Anna Lowenhaupt Tsing's 2005 study *Friction: An Ethnography of Global Connection*.[14] Tsing adopted the term "friction" in response to globalist narratives about unbridled movement, connectivity, and growth:

> The metaphor of friction suggested itself because of the popularity of stories of a new era of global motion in the 1990s. The flow of goods, ideas, money, and people would henceforth be pervasive and unimpeded. In this imagined global era, motion would proceed entirely without friction. By getting rid of national barriers and autocratic or protective state policies, everyone would have the freedom to travel everywhere. Indeed, motion itself would be experienced as self-actualization, and self-actualization without restraint would oil the machinery of the economy, science, and society. (2005: 5)

This fantasy of effortless expansion mirrors the philosophy of frictionlessness: both visions eulogize a world where connectivity and optimization are effortlessly pursued. To bolster such narratives, both capitalist globalization and the paradigm of frictionlessness need to conceal the venomous elements of their production (2005: 68). The lens of friction highlights the strained and conflicting realities elided by such narratives. While Tsing draws on friction as an umbrella term to describe the many precarious, provisional, and destructive practices required to uphold an image of unhindered global connectivity and circulation (2005: 4), I employ a more restricted definition: I mostly delimit my usage of friction to describe deviations from how a technology is normatively perceived. Friction arises whenever something

---

[14] Friction (or, rather, the Dutch term "frictie") is also the subject of researcher and essayist Miriam Rasch's recent book *Frictie* (2020). Rasch presents friction primarily as a counterweight to what she calls "dataism," or the belief that everything in the world can be made quantifiable and solvable through data. She endorses friction as indicative of those stimulating aspects of existence that slip through the cracks of datafication.

challenges the imperceptibility, efficiency, and smooth functionality of frictionless technology.[15] The notion of a hauntological aesthetic of imperfection, as discussed in the previous section, offers one way of introducing currents of friction into a technology's transparent, frictionless veneer.[16] More specifically, the friction that imperfection aesthetically engenders imbues the perception of the user with a sense of technological finitude and fragility.

Sean Cubitt's notion of *finite media* makes emphatic how media technologies are never exempt from the hauntings of finitude:

> Media are finite, in the sense both that, as matter, they are inevitably tied to physics, especially the dimension of time; and that their constituent elements—matter and energy, information and entropy, time and space, but especially the first pair—are finite resources in the closed system of planet Earth. Because they are finite, media not only cannot persist forever; they cannot proliferate without bounds. (2017a: 7)

Cubitt, by arguing that media are inextricably bound to the temporal, essentially appeals to the existential primacy of imperfection. In the previous chapter, I discussed how the Derridean notion of autoimmunity and the universal logic of imperfection that can be inferred from it spell an incurable condition of finitude and fragility. The risk of death is, therefore, always-already written into every conceivable machine. Cubitt's concept, however, also concerns the material manifestations of the existential condition of finitude. This dimension of finite media encompasses not only media

---

[15] In this context, the words of product designer Nick Babich, writing for Shopify.com, are revealing: "In experience design," he describes, "friction is anything that makes users stop and think" (2018).

[16] Any notion of an aesthetic of imperfection in a digital context can be embedded in a wider lineage of thinking about the concept of *post-digital aesthetics*. This concept, popularized by composer Kim Cascone around the turn of the twenty-first century in his seminal text "The Aesthetics of Failure" (2000), delineates an aesthetic logic that counters the "teleological movement toward 'perfect' representation" (Andrews 2002) and the rise of digital interfaces whose standardized lay-out encourages art that feels "too mechanically perfect" (Kelly 2009: 271). Arguably, then, the concept exhibits an aesthetic preoccupation with imperfection that neutralizes the alleged perfection of the digital; media theorist Florian Cramer maintains that "the simplest definition of 'post-digital' describes a media aesthetics which opposes [both] digital high-tech and high-fidelity cleanness" (Cramer 2015: 16). Cramer's words are paradigmatic of a recent renewal of academic interest in the concept of the post-digital (see Müller and Aich 2019; Berry and Dieter 2015a; Betancourt 2017; Bishop et al. 2016; Openshaw 2015; Contreras-Koterbay and Mirocha 2016), but this renewal is marked by a definitional dilution that diminishes the concept's relevance to my study (see also Kemper 2023). As such, I will not explicitly draw on it in my aesthetic analyses.

technologies themselves but also the worlds such technologies sustain and erode: technologies rely on matter, energy, and labor that are never infinitely exploitable. The philosophy of frictionlessness, as my pharmacological reading has underlined, comprises a materially destructive organization of technology that attempts precisely to rid perception of the recognition of medial finitude.

Cubitt implies that any challenge to the toxicities and finitudes that mark contemporary media must integrate the question of aesthetics (2017a: 151). As demonstrated, the aesthetic effects of frictionlessness reinforce rather than allay its poisonous tendencies. Imperfection, by contrast, forms a possible aesthetic principle through which to reconstrue perception in a more pharmacologically sustainable fashion. In the previous chapter, I have discussed the aesthetic category of imperfection as something that materially attunes and orients people to the fragile conditions of existence. By implication, aesthetic expressions of imperfection can, in the context of technology, register both the finitude and imperfection that must haunt all machines *and* the specific, material way in which these conditions unfold. As the notion of hauntological aesthetics further stresses, an aesthetic of imperfection can solicit an investment in technology's finite nature. This possible investment is a direct consequence of the human capacity to recognize the existential primacy of imperfection. I have suggested that this recognition enables us, but, because it denotes a capacity that technology itself lacks, also urges us to remain responsible for the shapes technology inhabits and for the material, ecological effects it engenders. As chronolibidinal beings—beings that are endowed with desire precisely because they are cognizant of loss—humans are capable of being aesthetically affected by conditions of finitude and can decide to temper destruction when they see it. Anna Wiener, in the account that opened this chapter, describes how Silicon Valley seeks to dissolve the "unnecessary friction of human behavior" and "decision-making" by imperceptibly subsuming all activity into technology (2020: 136). The arguments I have developed show, however, that the faculties of recognizing finitude, loss, and imperfection make human decision-making an indispensable element of human–technology interaction. In fact, in today's spectralizing times of frictionlessness, these qualities should be pharmacologically accentuated rather than nullified.

An aesthetic of imperfection provides one way of appealing to these human faculties: it can generate a sense of friction that heightens the feeling of finitude and fragility, thereby encouraging alternate engagements with technology. In the following three chapters, I analyze several digital objects that draw on an aesthetic of imperfection to examine the concrete ways in which the philosophy of frictionlessness and its core values can be countered.

Each of these objects intervenes into a different aspect of the perceptual and aesthetic experiences through which frictionlessness sustains itself. The defunct video game *GlitchHiker* (2011) that I investigate in Chapter 3 kindled empathy in its players and thereby forged an emotional attachment to the individual existence of a technological object, challenging the current devaluation of practices of maintenance and repair. The audiovisual performance *The Collapse of PAL* (2010) that I survey in Chapter 4 recounts a wider narrative that grieves the ruinous underside of technological innovation, gesturing not just at the finitude of individual objects but also at the larger cultural logic of condemning and discarding the outmoded. In Chapter 5, I examine the musical works of vaporwave producer 猫 シ Corp., and particularly his 2018 album *Palm Mall Mars*, as aesthetic reflections on how today's dominant forms of tertiary retention have constituted a culture of rampant consumerism. Each of these frictional objects suggests the possibility of a more care-inflected and sustainable conception of technology.

Care implies not only a meticulous attitude but also the fundamental faculty of having one's relation to the outside infused with feelings of concern, empathy, and responsibility. My study explores what it would mean if we were to include technologies in this outside. The objects I analyze draw on a technological and hauntological aesthetic of imperfection to elicit such a feeling of care for the realm of technology. They outline a model that, if generalized, would enact an alternate form of human–technology interaction that strays from the ghost-producing machinery that constitutes frictionlessness. These haunted artworks thereby respond to what I take Wendy Hui Kyong Chun to mean when she speaks of making our interfaces "more productively spectral" (2011: 60): while technologies will always bring forth ghosts, it is incumbent on us, as chronolibidinal and pharmacological beings, to assume responsibility for their shape and to account for the horizons of their haunting.

3

Silicon Ashes to Silicon Ashes, Digital Dust to Digital Dust

Chronolibido and Technological Finitude in *GlitchHiker*

In 2012, the notable music website *A Closer Listen* published an article on William Basinski's *The Disintegration Loops* (2002–3), a collection of albums (also discussed in Chapter 1) composed of the sounds of technological deterioration. In this article, tellingly titled "A Landscape of Decay," writer/musician Zachary Corsa presents Basinski to exemplify the rich tradition of experimental music—a tradition, Corsa explains, in which a positive value is ascribed to a technological aesthetic of imperfection and failure:

> No other visceral aesthetic is as aggressive in the message it broadcasts, that this technology is imperfect, as we're imperfect, that these mediums and these sound-waves are being pushed to their absolute limits and failing there, and there's beauty in that failure, beauty in the imperfection and flaws of man-made machinery, in a way that almost humanizes those flaws, makes them a character of the song as much as any melody, any run of notes. (2012)

Corsa clearly implies that the technological imperfections he admires harbor a resemblance to humankind's own flawed state ("this technology is imperfect, as we're imperfect"). His account suggests that a technology's aesthetic appeal rests in part on how its imperfections—understood here as notable flaws or failures in a technology's operations—resonate with our own fragile constitution. Effectively, Corsa intuits the workings of the existential primacy of imperfection I charted in Chapter 1; his words connote the constitutive impossibility, affecting both humans and their technologies, of being fully immune to finitude, alterity, and breakdown. Corsa, moreover, affirms my earlier argument that a technological aesthetic of imperfection can emotionally attune people to this universal condition and its material effects.

There is, however, also something problematic about Corsa's text. His impassioned rallying cry juxtaposes an appreciation for imperfection with a rather unrefined denunciation of digital technology. Corsa envisions a monolithic shift from analog to digital that has inaugurated an "era of looking sadly backwards, of revisionist history, when the dead obsolete technologies of our childhoods become crucial to sanely surviving in an Apple world that feels colder with every new sweatshop-fashioned model of iPhone" (2012). The technological perimeters that Corsa scours for the sweet scent of imperfection invariably turn out to be analog in nature; he advocates an "admiration for imperfection in a plastic surface world" that should lead us back "not just to cassette, but to scratchy Super 8 film, to blurry Polaroid and Holga photography, to tracking-damaged, pitch-warbling VHS" (2012). Corsa's critique pivots, in other words, on a stark opposition between the intimacy of the analog and the anemia of the digital. In Corsa's view, digital technology comes clad in a sterile shroud and its adverse effects can only be assuaged by turning to the warmer sanctuaries of a technological yesteryear.

Corsa's plea against the digital evokes multiple critical questions. Have people not become deeply entwined with the digital devices that comprise this so-called plastic surface world? Is digital technology itself, moreover, not markedly prone to failure? Do users not regularly experience dismay at the hands of malfunctioning phones or uncooperative Wi-Fi connections? It is the wager of this chapter that these minor tragedies of a purportedly frictionless digital culture, and the various responses they may elicit, are in fact not so minor at all, but rather have a leading role to play in the drama of a digitalized world. This perspective rejects contrived dichotomies of a "perfect," "cold" digital realm and an "imperfect," "human" analog world. Of course, this does not mean that Corsa's account is entirely invalid; in the previous chapter, I have gestured at the spectral residues of dispossession and destruction that inform today's dominant design philosophy of frictionlessness. Yet, following the philosophical tradition, carved out by Plato and expanded by Jacques Derrida and Bernard Stiegler, of thinking of technology as pharmacological in nature—as holding both curative and poisonous potentials—one must contend that a technology is never simply one monolithic construct, that digital technologies always accommodate pathways for more sustainable modes of engagement, and that the capacity of the digital to deteriorate perhaps provides one such avenue.

In accordance with these observations and the arguments developed in the previous chapter, my primary aim here is to further conceptualize technological instantiations of imperfection as elements of friction capable of challenging the digital's often frictionless veneer. I do so by first scrutinizing the concept of *glitch* and its pronounced associations with imperfection and spectrality.[1] This analysis is followed up by and embedded in a reading of *GlitchHiker* (2011), a video game, developed by the Dutch game studio Vlambeer, that tied its glitch-based aesthetic of imperfection to an overarching logic of technological finitude. *GlitchHiker* embodied a transient event that showed how feelings of care and empathy can emerge from the recalibration of technological norms and aesthetics. In order to account for the strong reactions the game elicited and to reflect on the wider cultural significance of emotional attachments to technological objects, I return to Martin Hägglund's earlier discussed concept of chronolibido. I conclude the chapter by demonstrating that reading *GlitchHiker*, as a digital object that lionizes finitude and fragility, in relation to current conditions of frictionlessness opens up new modes of thinking about human–technology interaction.

Glitch, Imperfection, Ghost in the Machine

In popular language, the term "glitch" refers to a perceptible moment of faulty interference in the routine operation of a (usually digital) technology. It indicates, in other words, the technological capacity to err and to decay. Glitch, then, is an agent of friction and attrition that negates any "assumed 'perfection' of digital technology" (Brøvig-Hanssen and Danielsen 2016: 98). It is an apparition whose possible advent the user has learned to apprehend— as Jacques Derrida insists, when engaging technology, one is always haunted by the "silent awareness [that one is] never safe from accidents, more common with the computer than with the typewriter or pen" (2005: 23). Glitch describes one such type of accident; generally short-lived, it creates a hyphenated user experience, a composition of error and correction.

The aesthetic effects associated with glitch—often taking the form of a "tumorous blob of digital distortion" (Manon and Temkin 2011)—have been repurposed in art, frequently with critical intent. This artistic lineage contains both a visual and an aural component. In the realm of music, glitch, while embedded in a longer tradition of soliciting the sounds of breakdown and malfunction, came into fruition as an actual genre in the 1990s. As media theorist Caleb Kelly describes, "glitch music combined the 'clean' world of the

---

[1] This discussion will also inform the next chapter, where I further assess glitch's critical capacities.

digital with a 'dirty', detritus-driven sound that switched the ratios of signal to noise in the realm of digital production" (2009: 8). On the visual side, film scholar Michael Betancourt traces the birth of glitch aesthetics back to the 1978 video *Digital TV Dinner* (produced by artists Jamie Fenton, Raul Zaritsky and Dick Ainsworth) that exploited the glitchy graphic effects that occurred when manually switching the game cartridge in the *Bally Astrocade* game console while the console was turned on (Betancourt 2017: 29–31). It would take some years, however, before the notion of visual glitch art fully entered art theory's vernacular—glitch theorist Rosa Menkman dates this moment back to around 2005 (2011b: 7). In both its aural and visual manifestations, glitch art highlights the failures, contingencies, and material compositions of technology.

This focus on material contingency calls to mind the aesthetic effects that Corsa contemplates, and it is thus not surprising that the concept of glitch and the technological fragilities it uncovers are often labeled in terms of imperfection. Take, for example, the following description of glitch's revelatory potential, offered by media critic Ed Halter: "The very moments that indicate the specificity of the medium occur when that medium starts to break down, to suffer and reveal *imperfections*. The technology becomes visible through its failures. Glitches and errors constitute evidence of its origins; we see the material through disruption" (2009: 71, emphasis added). Or observe the first point of Menkman's "Glitch Studies Manifesto," which warns us that the quest for noiseless transmission and perfect representation is a dogmatic fantasy that is doomed to remain precisely that: a fantasy (2011: 11).[2] Every medium, Menkman writes, carries its own "inherent fingerprints of *imperfection*" (2011: 11, emphasis added). Perhaps the most straightforward example of the discursive interweaving of glitch and imperfection is the title of a 2009 compendium that displays the work of numerous glitch artists: *Glitch: Designing Imperfection* (Moradi et al. 2009). The relation between glitch, technology, and imperfection is thus twofold: glitch signifies a particular aesthetic style often described as imperfect, and this aesthetic in turn reveals the perceived imperfections of the underlying technology. In this sense, glitch speaks to the sensibilities of hauntological aesthetics I addressed in Chapter 2: hauntological art tends to draw on an aesthetic of imperfection to probe the materiality and latent qualities of technology, and glitch marks one of its prime strategies.

The link to hauntology also discloses that the spectral tone distinguishing some of the terms used thus far ("apparition," "haunted") is not incidental:

---

[2] On the history of this quest for perfect mediation and sonic purity, and its ultimately misguided nature, see also Kromhout (2021).

one of the most common descriptions of glitch is that it is a "ghost in the machine"[3] (Goriunova 2012: 65), a "ghost" that "indicates an other subject in the medium" (Cubitt 2017b: 20). Considering my interest in hauntological aesthetics and the spectrality of technology, it is worth dissecting these spectral associations. What, exactly, is it that makes a glitch ghostly? For one, its time is profoundly out of joint. This is not only because it unsettles the regular rhythms of the medium, but also because it folds several temporalities within itself. Literary scholar Marc Olivier states that "[b]y definition, a glitch is an error that gets corrected" (2015: 261). As a result, something can be labeled a glitch only *after* it has come to pass, *after* a correction has taken place: "Like a ghost, [a glitch] is always already dead" (Olivier 2015: 262). It thereby signifies a spectral condensation of past, present, and future. Yet, glitch's spectral currents run deeper than this: glitch betrays three ghostly qualities that are especially relevant to the notion of hauntological aesthetics and the other forms of spectrality I have, in previous chapters, defined to outline a pharmacology of frictionlessness.

First, because glitch describes a functional aberration that may rip the user from her immersion, it can, like a ghost, shed light on norms that otherwise remain unseen, elucidating practices that have become so habitual that they evade conscious awareness. Ghosts, as sociologist Avery Gordon has suggested, destabilize "the propriety and property lines that delimit a zone of activity and knowledge" (1997: 63). Glitches potentially trigger exactly such an epistemological subversion—as glitch theorists Hugh S. Manon and Daniel Temkin describe: "[w]hether its cause is intentional or accidental, a glitch flamboyantly undoes the communications platforms that we, as subjects of digital culture, both rely on and take for granted" (2011). Manon and Temkin's words are symptomatic of a wider trend among glitch theorists to ascribe to glitches a pedagogical, sometimes almost gnostic power.[4] This power derives from glitch's foregrounding of failure, which, in diverging from technology's routine operations, is often theorized to guide its audience to new and emancipatory insights about their reliance on technology. Such

---

[3] The notion of the ghost in the machine comes from philosopher Gilbert Ryle's 1949 work *The Concept of Mind*, where he criticizes Cartesian Dualism, or the supposedly fundamental distinction between the mind and the body, as the "dogma of the Ghost in the Machine" ([1949] 2009: 5), the metaphor of the ghost here pointing to something vaporous that defies definition. In popular discourse, the description has since been used to refer to phenomena that appear to be affected by something that escapes perception or that exceeds neat systematization.

[4] This pedagogical quality is best exemplified by media historian Peter Krapp's suggestion that glitch-based aesthetics help to see that "what one needs to learn from mistakes is not to avoid them but something else altogether: to allow for them; to allow room for error" (2011: 92).

arguments place glitch theory firmly in the tradition of a Heideggerian phenomenology of technology; Martin Heidegger advanced the now canonical notion that to encounter a tool in a damaged or broken state is also to regard it in a different and potentially more reflective light ([1927] 2010: 102–4). This observation can be extended even further by drawing on Stiegler's postphenomenological perspective and his framing of technology as an essential element of human consciousness (see Chapter 1). Following this perspective, glitches can remind one that the very acts of perceiving and cognizing are utterly dependent on technical mediation. A glitch, by implication, potentially illuminates the fragile nature of consciousness, whose relation to itself and its outside is never given definitively, always susceptible to being rewritten and affected by equally fragile technological traces.

Glitch's elucidative capacities appear all the more pertinent in the age of frictionlessness, wherein "[m]ost computing power is used in an attempt to make people forget about computers" (Lialina and Espenschied 2009: 9). As the previous chapter elaborated, frictionlessness, appealing to the values of user-friendliness, connectivity, and optimization, aims to tie technology so seamlessly to the navigation of everyday life that technology's presence slips largely into the subliminal. Frictionless technologies are functionally designed to make as little infringements on consciousness as possible even as they facilitate and influence an ever-greater degree of tasks. The ghostly ability of glitch to bring the mediated nature of perception into sharp relief and to unravel the norms that so deeply determine our relation to our machines takes on new relevance in view of this situation. If frictionlessness aims to establish digital technology as a generalized unconscious norm that mediates all human activity, glitch's aesthetic power to expose processes of mediation poses a compelling counterweight.

At the same time, glitch's critical impact should not be presumed: the theoretical tendency to indiscriminately portray glitch as a causeway to enlightened experience has rightfully received criticism. Michael Betancourt states that "the problem for critical media is not the creation of stoppages, but the adaptability of the audience to the stoppage itself" (2017: 127), implying that glitches are not always the formidable specters they are made out to be. Digital technology, so Betancourt suggests, trains its users to disregard any abeyance as an inconsequentiality that has no direct bearing on the "perfect" image of the digital itself (2017: 102). Moreover, as Chapter 2's discussion of the dominant value of optimization suggested, technological imperfection may be a conscious design feature integrated in purview of the phenomenon of planned obsolescence. In this perspective, an increase in glitches ideally serves as the impetus for purchasing a new, more frictionless alternative and glitch may thus in fact be instrumental to solidifying digital norms of updating and replacement.[5] These observations imply that a capacity for subversion is no innate quality of glitch—as a technological ghost, its critical capacities are rooted in whether its arrival is acknowledged by the beholder and in the range of responses to which such acknowledgment is likely to give way. When is a glitch a liminal ghost that does not even pass the threshold of perception, when is it a dreaded specter that urges the consumption of the new, and when is it a ghostly force that impels an entirely new relation to its technological host? This question testifies to the importance of the notion of friction in discussions of glitch: a glitch's ability to affect the user tends to coincide with the forms of friction it manages to generate in user experience. The glitch-based objects I analyze in this chapter and in Chapter 4 effectuate a critical response precisely because of the friction they provoke.

A second, closely related ghostly quality of glitch is its capacity to reveal spectral and ungovernable forms of technological agency. The previous chapter explored a dominant sense of the spectral that was described as "technological spectrality," a term that defines the tendency of technology to evoke spectral associations when it produces material effects while remaining insensible or when its agency transcends that of the human. Frictionlessness not only conceals its own operations but also renders them increasingly far-reaching and complex, and the cognitive purchase the human mind has on these procedures is declining. Glitch is an aesthetic reminder not only of the fact of mediation but also of the complexity and contingency of technological operations themselves. Especially when they occur unexpectedly, glitches grant a brief glimpse of the spectral realms that hazily support the interface's frictionless facade: vitiating illusions of full control, glitches aesthetically disclose that intractable forms of technological agency may always arise unbidden, be it in the form of noisy interference, technical malfunction, or the imperceptible mediations of what media theorist Mark B. N. Hansen has termed "twenty-first-century media" (2015: 37).[6] In this regard, glitches

---

[5] Moreover, as media theorist Susanna Paasonen notes on the basis of essays that her students wrote, glitches thwart the internalized demand for speed and instant connectivity that frictionless technologies have wrought and thus frequently provoke frustration: "Lags, delays, and glitches interfere with [the] ideal of frictionless use, forcing a discontinuous rhythm to one's interactions and occasionally giving rise to sharp ripples of irritation" (2021: 35).

[6] This concept, which denotes the tendency for digital media to evade human perception and consciousness altogether, has been addressed in Chapter 2. There is a fundamental tension between glitch and twenty-first-century media: while twenty-first-century media are defined by their imperceptibility, a glitch, in order to be labeled as such, requires a perceptible interface to communicate itself to the user, often through an aesthetic of imperfection. Glitch, therefore, can never reveal the operations of twenty-first-century media directly, and Hansen would perhaps argue that glitch is a relic of modes of thinking that still frame human consciousness as the nexus of all media operations. I

function as mediums, those haunted figures that serve as intermediaries between seemingly irreconcilable realms (those of the living and the dead, the seen and the unseen, the earthly and the supernatural).[7] Glitch locates a principle of indeterminacy and contingency in the digital's kernel: by virtue of its ghostly ability to appear (ostensibly) autonomously and to reveal unruly powers, it transmits an ultimate "sense of the wilderness within the computer" (Manon and Temkin 2011) and underscores the digital's necessary backdrop of "ontological complexity [and] chaotic incompleteness" (Parisi 2013: 12; see also Marenko 2015).

Lastly, glitch, speaking to one of the core constituents of the spectral, is contiguous to death: it expresses the fundamental fragility of all technology. As the Derridean concept of autoimmunity—the inescapable condition of being threatened from within that was discussed in Chapter 1—spells out, there is no technology that is exempt from the spacing of time and from the violent threat of finitude that this announces. No technology, in other words, can be fully immune to breakdown and contingency, because these risks are written into the machine from its inception; when engaging a technology, one necessarily engages a fallible and finite apparatus. Glitch, as an aesthetic of imperfection that highlights failure and decay, can remind the user of this foundational element of fragility and therefore, like a ghost, stands in for death. As this chapter will demonstrate, glitches may thus serve to compose a form of chronolibidinal aesthetics, drawing out emotional attachments by alerting their audience to the possibility of technological loss. By reminding the user of the porous and terminable aspects of technology, glitch forms a ghostly note of friction that dismantles fantasies of fully frictionless or lossless technology.

Glitch may spell the individual demise of a technological host, but it also relates to conditions of imperfection and finitude in a broader sense. While glitches are dependent on human perception, the logic that informs them precedes and exceeds the human. As media theorist Sean Cubitt writes, glitch "belongs to the prehuman, inhuman universe against which we drag our messages into existence, and against which we strive to retain their integrity" (2017b: 22). Eroding whatever fleeting refuges of order that people craft from the vortices of matter, thought and media, glitch discloses an existential primacy of imperfection in which alterity and finitude haunt all that exists.

---

believe, however, that glitch retains a primary function for its medium-like capacity to temporarily disclose such technological otherworlds, marking a fleeting point of connection between the scrutable and the inscrutable. See also Kemper and Kolkman (2019); Kemper (2023).

[7] On mediums and their associations with the spectral, see also Peeren (2014: 110–43).

Integrity and stability are not preconditions that subsequently accede to glitch's corrosive grasp; glitch, on the contrary, renders transiently sensible how an existential logic of imperfection pervades all order and stability. Incapacitating though this may seem, glitch in fact attests to the ever-present possibility of change; following the notion of hauntological aesthetics, there are few technological agents that communicate as directly that technologies can always be modified and that technologically entrenched norms can always be denounced and redirected. Art theorist Curt Cloninger indeed finds a subversive agency in how glitch challenges "the ideal of a pure signal," an ideal that expresses a "metaphysical/Platonic attempt to downplay the immanent and maintain (the myth of) the pure transcendent" (2011: 35). Cloninger states that "subverting (literally 'deconstructing,' in Derrida's original sense)" such metaphysical beliefs in the possibility of technologically attainable transcendence and perfection "is a radical (root level) 'political' act" (2011: 35). Hence, glitch can be figured not simply as a ghost that resides in a particular machine but also as an inescapable possibility that exposes spectrality to be a universal condition: all things are haunted by finitude and imperfection, and therefore no pure, perfect, or infinite channel could ever exist.

The description of glitch as a ghost in the machine is, in short, a heterogeneous label that invites numerous spectral valences. In order to grasp the specific significance of glitch in the face of the philosophy of frictionlessness, I now turn to this chapter's case study, Vlambeer's video game *GlitchHiker*. This object, as I will demonstrate, tied its glitches to a conceptual logic of finitude and thereby construed a user experience that was more emotionally oriented to technological conditions of impermanence. Of the ghostly qualities I have just described, the capacities of glitch to presage technological death and challenge aesthetic conventions are especially pertinent in the context of *GlitchHiker*.

GlitchHiker

Vlambeer (2010–20) was a Dutch independent game studio, based in the city of Utrecht, that garnered international acclaim with several well-received titles (the most prominent of these being the 2014 shooter game *LUFTRAUSERS*). The studio was nested within a national culture that is globally renowned for its contributions to the gaming industry and that boasts a remarkably strong scene of independent developers, all facilitated by a supportive government climate (Roso 2013). In 2011, Vlambeer took part in the Dutch edition of the Global Game Jam. Game jams are events that gather people from a variety of video game development-related backgrounds and

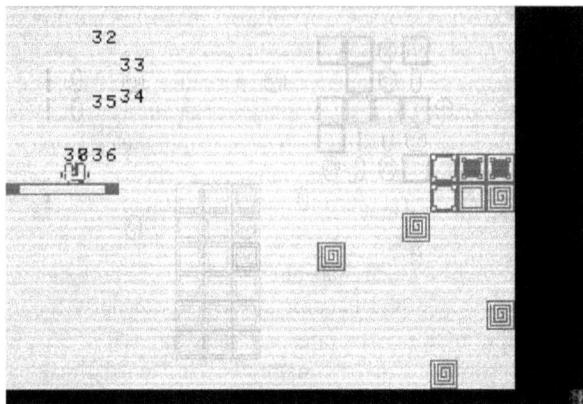

Figure 3.1 *GlitchHiker*'s glitch-based aesthetic. Image source: www.youtube
.com/watch?v=r2MDgKGCF88. Accessed October 31, 2022.

that are organized with a spirit of creativity and a willingness to experiment
in mind. Generally, the goal is to assemble a team of programmers, designers,
and artists, and to develop a game within a strictly limited amount of time.
During the 2011 event, Vlambeer coordinated a team of developers and
designers with the task of creating from scratch, within a 48-hour time slot, a
game that conformed to the event's theme of "extinction."[8] The result of this
challenge was *GlitchHiker*. The game relied heavily on glitch-based visuals
(Figure 3.1)—it would occasionally freeze, polychromatic schemes would
often irrupt into the screen, and the game's already pixelated aesthetic was
regularly distorted (Figure 3.2)—and on an adaptive score, composed by
Rutger Muller, that was itself accentuated by audible glitches and flaws.

What was most remarkable about *GlitchHiker*'s aesthetic is the fact
that its glitches signified the game's approaching demise. *GlitchHiker*'s key
gameplay element revolved around its life reserve. As is not uncommon in
video games, the player was allotted a finite number of lives, but what was
unique about *GlitchHiker* is that the game's code was programmed to erase
itself should a player drain the last life from the reservoir. After this final life
was expended, nothing more could be done; the game was terminated, never
to be played again. Its creators had hidden all its code behind a randomly
generated password, meaning that not even they could ever conceivably
retrieve it (Flores 2011). The game thus aesthetically and functionally
enacted a condition of autoimmunity: its existence was premised entirely on

[8]   In addition to Vlambeer's own Rami Ismail and Jan Willem Nijman, the team consisted
      of Laurens de Gier, Jonathan Barbosa Dijkstra, Rutger Muller, and Paul Veer.

Figure 3.2 *GlitchHiker*'s glitch-based aesthetic. Image source: www.youtube .com/watch?v=r2MDgKGCF88. Accessed October 31, 2022.

a capacity for self-destruction, death arriving not as an external force but as a risk already contained within the game's design. *GlitchHiker* went on to receive the first-place award in the Game Jam's competition from both jury and audience—a testament that the game and its idiosyncratic concept managed to strike a sonorous chord.

GlitchHiker, it must be noted, was certainly not the first of its kind: not in its evocation of glitch-based destruction, not in its technological evanescence and material self-erasure, and not in its glitchy reconfiguration of the medium of video games. A pioneering example of the phenomenon of glitch-induced ruination is found in the live performances of art group 5VOLTCORE. These performances were centered around a computer that the group proceeded to undermine through power interruptions and short-circuiting: "This process tortures the machine and makes it scream out shreds of powerfully colored images, until the computer eventually dies, which ends the performance" (Menkman 2011: 37). *GlitchHiker*'s auto-erasure finds antecedence in Dennis Ashbaugh, William Gibson, and Kevin Begos Jr.'s famed *Agrippa (A Book of the Dead)* (1992). This poem, published in book format and on floppy disk, destroyed itself after first use. More generally, *GlitchHiker* brings a twist to the genre of permadeath games—games in which in-game death results in the irrevocable loss of one's character (Chang 2019: 207). Lastly, an example of a similarly glitch-based video game is art collective JODI's *Untitled Game* (1996–2001), a collection of modifications of the influential video game *Quake* (1996) that challenges the hegemonic and implicit norms behind video game design by erecting glitchy reimaginations of the game-world. *GlitchHiker* similarly shed light on the norms of game design, strapping the medium's fixation on death to the underlying code. Here, however, the process of disintegration was conducted by the player and not by the artist.

Before further addressing *GlitchHiker*'s specifics, the game's finite functionality warrants a methodological consideration. *GlitchHiker* was a digital object programmed for extinction and this is precisely how things played out—soon after the Global Game Jam ended, a player wasted the last remaining life and no playable version of the game thus exists today (Polson 2011). The game was more of an event than an integrally archivable object. As such, in my analysis of the game, I am relying mostly on textual documentation[9] and on the sparse number of available videos that depict gameplay footage.[10] Problematic as this would be if I were to pursue an exhaustive textual analysis, I suggest that this is not an impediment in the case of the present inquiry, as I am predominantly interested in *GlitchHiker*'s constitutive logic of finitude. I am, to put this more concretely, mainly concerned with the game's concept and the way in which this concept was communicated to the player through a glitch-based aesthetic of imperfection. As this concept and its aesthetic visualization are clear from the existing documentation, no playable object is required—it is, on the contrary, precisely the fact that no playable version exists that is of prime importance.

The game's life system also deserves some further elaboration. *GlitchHiker* functioned according to a points-based system where, for every 100 collected coins, a life would be added to the life reserve, whereas anything below that threshold would result in the detraction of a life. While, in theory, it was entirely within the realms of possibility that the game would have remained preserved (provided enough players amassed sufficient points), its open-ended availability to an uninitiated crowd—cognizant of the stakes of play but unversed and inexperienced when it came to gameplay mechanics—practically ensured that the game would at some point meet its demise. *GlitchHiker*'s extinction-inspired concept thereby offered a frictional experience wholly unique to video game culture. While we are used to video games that abound in death on the level of gameplay, video games tend to keep this morbid infatuation away from the game's performance; dying in the game has no impact on the technology's capacity to operate. In the case of *GlitchHiker*, conversely, players were given a sense of responsibility over the well-being of the technological object itself; missteps did not merely affect

---

[9] In a further twist of the inevitable fate of technological finitude, many of these webpages have gone corrupt since the time I first started work on this chapter and can now be accessed only through the Internet Archive's Wayback Machine.

[10] Gameplay footage of GlitchHiker can, for example, be found at: www.youtube.com/watch?v=r2MDgKGCF88 and www.youtube.com/watch?v=P1h95LNDjw8.

the fate of the game's virtual protagonist but carried real consequences for the game's existence. Contrary to the familiar rationale of simply reloading and respawning when one's digital lives are wasted, in *GlitchHiker* one slipup too many would cause the game to be lost irrevocably.

In terms of its aesthetic design, *GlitchHiker* comes across as an ordinary indie game, opting for a mode of gameplay and a top-down viewpoint that recall an iconic series like *Bomberman*—a strategic video game where players must use bombs to find their way through mazes. *GlitchHiker's* gameplay dynamics were of a similarly straightforward and arcade-inspired variety, having players navigate a "single screen arena" where they would have to collect coins while avoiding obstacles (Kirn 2011). Its simplicity served to lend an optimal degree of poignancy to the game's macabre concept. This concept was fortified by an aesthetic that used glitches to reflect the game's stamina, with the music gradually disintegrating and the game visibly struggling to stay operative as the life pool slowly emptied. Journalist Jeremy Peel, writing for online magazine *PCGamesN*, describes how, "as the number of lives in the pool inevitably began to drop, the game's health visibly and audibly deteriorated. Its sickness became increasingly evident in the glitches that obscured parts of the screen, and at those times in which the action froze entirely; the game would hang for seconds, before lurching horribly back into life" (2014) (Figure 3.3). Not only, then, did *GlitchHiker's* glitches counter the aesthetic qualities of frictionlessness (challenging the smooth transmission

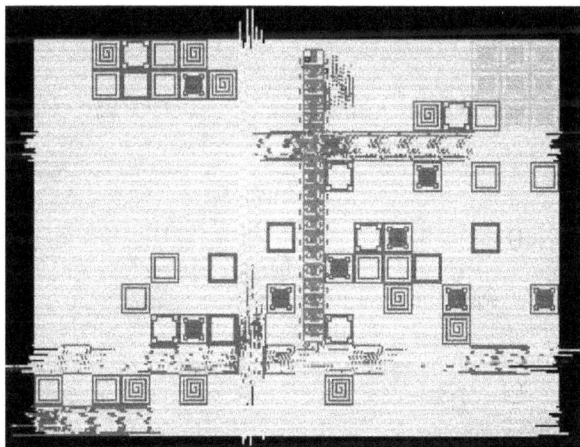

Figure 3.3 *GlitchHiker's* glitch-based aesthetic. Image source: www .gamasutra.com/db_area/images/igf/GlitchHiker/screenshot.jpg. Accessed April 26, 2021.

of information and the transparency of mediation), but they also greatly amplified glitch's spectral capacity to signify technological breakdown and death. Moreover, because the game's glitches directly impacted gameplay experience, they generated an unignorable sensation of friction. This friction had the ultimate function of keeping finitude and fragility at the forefront of perception.[11]

The phenomenon of a fragile technology floodlighting its inherent finitude prompted players to experience feelings of remorse or even to flat out refuse to play the game for fear of contributing to its conspicuous agony and ultimate expiration. Reflecting on what stood out most about the *GlitchHiker* experience in the months following its passing, co-creator Rami Ismail indicates that it is "empathy": "The story of *GlitchHiker* wasn't so much in the game, as it was a thing happening to the players. That breeds the circumstances in which responsibility can exist—in which guilt can exist— and in which such emotional attachment can happen" (quoted in Peel 2014). In another interview, Ismail reflects more extensively on the emotional response the game elicited:

> People were actually empathic towards the game. The moment the first person said, "I'm not playing this, I could kill it!" was the moment we realized we made something unique. . . . It all seems playable and fun, but the moment you sit down you're confronted with the reality that what you do could potentially withhold that experience from everyone else, forever. Some sat down and decided they couldn't do it. Some were scared. Those that scored new lives tried to score more. Those that lost lives felt guilt and scurried away. (quoted in Polson 2011)

GlitchHiker's concept and its aesthetic reinforcement thus generated a notable sense of friction in the audience; *GlitchHiker* "was a thing happening to the players" (Ismail quoted in Peel 2014) that evoked a range of responses—from a reluctance to take on responsibility to a diligent effort to keep the game alive. These responses, though divergent, all betrayed a sense of empathy and care, which leads journalist Matthijs Dierckx, writing for *Control*, to suggest that the best way to understand *GlitchHiker* is to view it not as a product designed for fun and play but rather as a work of art (2011), a digital object that created the conditions for a unique emotional connection to arise. As Ismail avers, *GlitchHiker* counts as "proof that there's potential to reach out to people's emotions beyond the typical game emotions (such as stress

---

[11] While I have focused primarily on GlitchHiker's visual instantiations of glitch, a similar argument can be made about the game's score: as the game approached death, the music was increasingly characterized by distortions and sonic lapses.

or excitement) through game mechanics instead of narrative techniques" (quoted in Polson 2011); *GlitchHiker* probed how unconventional feelings of care and responsibility can emerge from the recalibration of familiar vectors of human–technology interaction.[12]

GlitchHiker, to sum up, was bound to finitude in two ways: finitude defined the game's concept, and the sense of this irrevocable finitude was subsequently heightened through the ghostly imperfections of glitch. This redoubles, within the small scale of the game's universe, the relation between the existential logic of imperfection and the aesthetic of imperfection as examined in Chapter 1. *GlitchHiker*'s entire game-world was defined by a primacy of imperfection: it enacted a condition of autoimmunity that could not be transcended or redeemed because finitude was inherent to its very being. In the material presentation of the game, a feeling for this irrefutable finitude was aesthetically established and intensified through a spectral and glitch-based aesthetic of imperfection that underlined breakdown and loss. The game thus coupled a constitutive technological finitude with a frictional aesthetic of imperfection to encourage a sense of care for technology. *GlitchHiker* thereby provides a model for thinking about how the human capacity for care, informed as it is by a recognition of impermanence, can be magnetized by technological objects that heighten a sense of finitude within perception, and about what the cultural significance of such forms of care might be in times of frictionlessness.

GlitchHiker, Care, and Chronolibido

As a work of glitch art, *GlitchHiker* provides a profound case to support technology journalist Chris Baraniuk's suggestion that "[g]litch art is just the beginning of our culture leaning towards a world in which the permanence of the digital is no longer assumed" (2013). Likewise, the game strongly adheres to what art theorist Anna Munster, in a Stiegler-inspired analysis, describes as an emergent aesthetic "preoccupation with digital death" (2011: 68). *GlitchHiker* indeed comprises an object that is "cognizant of finitude, consequence and even death" (Munster 2011: 69) by dint of the aesthetically intensified prospect of its extinction. Instead of abiding by video games' usual logic of presenting "the digital as merely an opportunity to inconsequentially

---

[12] The wider capacity for video games to elicit empathy is increasingly drawing scholarly attention (see, for example, Muriel and Crawford 2018: 123–30), but this empathy is seldom theorized as extending to the video game as a material object itself.

'reload' and refire" (Munster 2011: 70), it caused the actions of players to carry material consequences by incorporating the risk of irreversible medial loss. Akin to Baraniuk, Munster suggests the growing artistic interest in the digital as a space that accommodates plenty of death to serve as a correction to paradigms that exclusively equate the digital with endurance, connectivity, and vitality. I propose, however, that the main cultural issue that *GlitchHiker* potentially speaks to is not so much that digital impermanence is denied or repressed, but rather that it tends to be assimilated within a destructive, profit-driven dance of obsolescence and renewal.

I posit, more exactly, that *GlitchHiker* is indicative of a *different* way of anticipating technological transience: of perceiving digital devices not as entities whose life span is to be curtailed through substitution and wastefulness, but as unique and individual objects whose implications of finitude may enliven and make more committed one's investment in them. The former rationale characterizes the philosophy of frictionlessness, with its fetishization of relentless optimization, and encourages a relation to digital technology that sees in its fallible nature an incentive to discard rather than tend to objects when they first show signs of imperfection and finitude. Frictionlessness stimulates a deep investment in the perpetuation of its digital logic, but a relatively evanescent, improvident, and ultimately careless relation to its individual devices. *GlitchHiker*, on the contrary, sparked an attachment to its own unique material existence. The implications of this phenomenon can be better understood by reading Martin Hägglund's concept of chronolibido in relation to science and technology scholar Steven Jackson's work on repair and maintenance.

I have discussed Hägglund's concept of chronolibido at length in Chapter 1, but it is worth briefly reintroducing it here. Chronolibido discloses how temporality and the attendant possibility of loss are an inherent part of the very structure of care and desire. Hägglund demonstrates that the fact that an object can be lost—grievous though the promise of bereavement may seem—forms the condition for any affective attachment to assume shape in the first place. All forms of attachment are necessarily informed by a constitutive recognition of and investment in temporal finitude, which marks the wellspring of both prosperity and calamity. Hägglund develops his arguments in reference to the Derridean notion of the spacing of time: because of the structure of the trace, which can retain a moment for the future only by exposing itself to that very same future's uncertain nature, to be granted the possibility of living on is equally to be exposed to the risk of harm (2012: 16). This also applies to the objects of our attachment: to be invested in their survival is necessarily to be invested in their finite nature and possible destruction. As Hägglund summarizes, "care in general" depends on the "double bind" of finitude (2012: 9). This double bind designates that to be bound to pleasure is to be bound to pain, to be bound to love is to be bound to grief, to care for something is at the same time to admit the possibility of its dissolution—one cannot erase one half of the bind while preserving the other.

It is important to reiterate that chronolibido, though central to temporal perception, is insufficient to effectuate a particular response. However, in Chapter 1, I have discussed how Hägglund's arguments can be augmented by an aesthetic perspective that queries how representations of the temporal can intensify the flow of chronolibido. I have furthermore argued that an aesthetic of imperfection can play an affirmatory role in evoking chronolibido, indicative as imperfection is of the passing of time and of the ultimate impossibility of purity and permanence. By drawing on the work of Stiegler, I have, moreover, explained how technologies ground and mediate the human perception of time, and how they are therefore enmeshed with the channeling of chronolibido. The way technology phenomenologically presents us with temporality and finitude impacts how we act and respond. As an example, one may think back to the approach of hauntological aesthetics addressed in Chapter 2: hauntology draws on a technological aesthetic of imperfection to chart the intersections of perception, memory, and anticipation that technology facilitates. Hauntology's appeal stems in large part from how its aesthetic of imperfection dramatizes temporality, technological corruptibility, and finitude, and from how it thereby quickens the audience's chronolibidinal investments.

GlitchHiker and its overarching ethos of digital death show that technologies can themselves become the object of chronolibidinal investments. Because the game was premised on a mechanical dramatization of the logic of autoimmunity and because its glitch-based ghosts so clearly communicated approximate decease, players came to care for the game in a fashion that diverges from more functionalist modes of engaging digital objects. The feelings of care, guilt, and responsibility that *GlitchHiker* elicited were caused, in other words, by how emphatically it affirmed the unicity and fragility of its existence. For many players, playing the game thus meant taking on the responsibility, strongly affected by the flow of chronolibido, of aiding in its survival by expanding its life reserve. Because survival is never guaranteed, living on and keeping faith with what one cares for generally requires recurring acts of sustenance, attention, and restoration. *GlitchHiker* was no exception to this rule: it offered a mode of play that essentially consisted of nothing more than repair and maintenance work. Its chronolibidinal aesthetic urged players to manage emergent glitches and to ward off impending death, and it thereby allocated a central rather than liminal role to negentropic acts of repair and preservation. The players' restorative acts could, however, never grant *GlitchHiker* full immunity—engaging it would always mean engaging a fragile, autoimmune object for whose well-being and possible extinction one was made accountable.

The cultural significance of the sense of care that *GlitchHiker* solicited can be clarified by examining Steven Jackson's account of repair and maintenance. Jackson calls for a more serious consideration of just how central practices of repair and maintenance must be to the contemporary mediascape: "[B]reakdown, maintenance, and repair constitute crucial but vastly understudied sites or moments within the worlds of new media and technology today" (2014: 226). As swaths of energy are expended to kindle the flames of optimization, connectivity, and innovation, the banality of the common glitch indeed reveals that the technological infrastructures that facilitate everyday life are in a constant state of decomposition. Against an industry logic of seamless and frictionless technology, propelled by a destructive drive for perfection, the attitude of repair takes seriously the individual finitude and fragility of technology. Jackson advocates a mode of "broken world thinking" (Jackson 2014: 221), a form of apprehending technology that departs from the primacy of dust and decay instead of privileging productivity and progress. Implicitly, Jackson accentuates the value of chronolibido as a faculty that can inflect the human relation to technology: repair and maintenance derive from the recognition of an object's finite nature and from a desire to keep its oblivion at bay rather than acquiesce to it. Focalizing technology through a lens slanted toward moments of breakdown and disarray discloses, moreover, that practices of repair and maintenance are the spectralized fuel that allows illusions of stability and permanence to blossom.[13] Nonetheless, the prime value of optimization negates such practices: it stimulates a consumption of the new and discourages the conscientious work of remaining with the fragile and the outmoded.

Glitches, in their ghostly signification of technological finitude, may provoke a wealth of different responses: from indifference or the replacement of what is broken to a commitment to repair and restoration. As stressed, this is always dependent on contextual factors, and *GlitchHiker*'s glitches and the friction they produced turned out to sustain an environment that primarily inspired empathy and care. Significantly, Jackson repeatedly invokes the term "care" in relation to repair and maintenance, not only to capture the multitude

---

[13] This spectralization of (technological) practices of maintenance and repair, also mentioned in the previous chapter, can be read in relation to a longer, race-, gender- and class-informed history of marginalizing and minimizing reparative, domestic, and reproductive labor—see, for example, Duffy (2017); D'Ignazio and Klein (2020).

of activities required to ensure that everyday technologies remain more or less functional, but also to indicate a possible moral and political aspect to how people relate to technology: "[T]he ethics of repair admits of a possibility denied or forgotten by both the crude functionalism of the technology field and a more traditionally humanist ethics (which has mostly ignored technology anyway). What if we care about our technologies, and do so in more than a trivial way?" (2014: 232). This question broaches a speculative angle that allows for a conception of alternate futures: what, we might ask with Jackson, would more care bring to the pharmacological world of technology? And how might technologies themselves modulate such bonds of care? Could we, by implication, conceive of a different chronolibidinal economy of technology that inspires less poisonous and transitory objectual and material attachments?

GlitchHiker provides a tentative answer to these questions and accordingly comprises a possible model for thinking through Stiegler's call for the reconstruction of a "libidinal economy (a philia), without which no city, or democracy, or industrial economy, or spiritual economy, is possible" (2011b: 15). In Stiegler's view, such a libidinal economy binds ecology, technology, and political economy (Stiegler 2017a: 137) and incorporates as *pharmaka* the technological objects that constitute the technical milieu, necessarily involving the forms of care these objects encourage or impede. As I have argued earlier, pharmacological perspectives must take a holistic view of technology and question not only how the individual mind is (toxically or tonically) affected by the *pharmakon* but also how technological modes of production and consumption stimulate their users to relate to and take care (or not take care) of their environment: of themselves, but also of each other, of the ecological realm and of the *pharmaka* that co-constitute them. Significantly, Stiegler's plea is spurred by what he views as the destructive tendencies of the digital, whose devices disproportionately aim to reroute "libido . . . towards the interest of consumption" (2014: 21). Indeed, frictionlessness relentlessly promotes the consumption of new appliances and updates, and discourages sustained temporalities of care, or, more pointedly, care for the realm of technological objects. *GlitchHiker*, by contrast, evoked the cadence of an alternate temporality, one marked by a protracted investment in its singular technological existence.

GlitchHiker, however, also highlights a limit in Stiegler's conception of a new libidinal economy of technology, because Stiegler ultimately negates the way in which the element of *chronos*, as Hägglund has demonstrated, constitutes care and desire. Stiegler embeds his vision of a reconstrued libidinal economy in a Lacanian tradition, thereby figuring desire as formed through an image of infinity and transcendency: "Desire," Stiegler posits, "is structurally related to infinity" (2009a: 47). He locates a "point of singularity at the origin of all deconstruction" that is informed by "the libidinal economy of an infinite desire for an infinite singularity on the part of a singularity itself infinite" (Stiegler 2013: 43). Hence, Stiegler believes that without "the question of the infinite" there can be no desire and no care (2019: 323): the possibility of caring about something is for him mediated by an unconscious that operates on a projection of and aspiration for the infinite, for a perfect state of being that would be unmarked by temporality.

Hägglund's concept of chronolibido shows, however, that care and desire are emphatically not premised on infinity, but rather on a fundamental attachment to the finite. To reiterate, rather than care having as its "repressed truth" a yearning to overcome the ontological deficit that condemns us to temporality, care is formed *only* through a constitutive investment in finitude (Hägglund 2012: 145). A libidinal economy always derives from a "being bound to the mutable and losable, which is a condition for libidinal being in general" (Hägglund 2012: 123). Basing the image of a reconstrued economy of care on a notion of infinitude misperceives the existential primacy of imperfection and will thereby also fail to grasp the full implications of imperfection in its aesthetic manifestations. *GlitchHiker*, as an object whose irrevocable finitude formed an immediate source of care, makes these implications tangible. The game provided a dramatic manifestation of how, as Hägglund insists, it is "the temporal finitude of the catchected object [that] calls forth the economic capacity to redistribute resources or withdraw investments as a strategic response to being dependent on what may change or be lost" (2012: 113). *GlitchHiker* indeed showed the ambivalences to which finitude may give rise—some individuals took on responsibility for its survival, others opted to move on—but the overarching point is that these attachments and withdrawals were a direct response to *GlitchHiker*'s transitory nature. *GlitchHiker* concretized a technological, *chrono*libidinal economy that aesthetically kept loss and fragility at the forefront of perception, making the ramifications of technological impermanence more palpable. It thus encouraged, in many of its players, a deeply material attachment to technology as a finite and vulnerable object.

Extinction

Contrary to the words of Zachary Corsa that opened this chapter, digital technologies, like the analog technologies that precede them, are notably susceptible to breakdown, contingency, and decay. This chapter has shown how this susceptibility can be aesthetically and hauntologically appropriated to provoke alternate conceptions of technology. Because care and desire are informed by the anticipation of death, technologies can potentially deepen one's investment in and relation to them by formalizing and aestheticizing the condition of their finitude. *GlitchHiker* actualized how glitch, as a frictional ghost in the machine that portends demise, can serve as a prime instantiation of this notion. Through its palpable concept and its intensification of glitch's ghostly ties with death, the game provided a counterweight both to illusory temporalities of (digital) infinitude and to the destructive temporalities of obsolescence that define frictionlessness: while the game's obsolescence was certainly planned, this was a design devised to absorb finitude more meaningfully into gameplay experience.

GlitchHiker and its aesthetic of imperfection, while modest in scope, offer an empirical example of how the condition of chronolibido can inflect a more sustainable relation to individual technology. As *GlitchHiker*'s co-creator Rami Ismail deduces, the game proved that rewriting game design's normative mechanics can open up avenues for different perceptions of technology (quoted in Polson 2011). Of course, considering the game's minimal reach, it would go too far to present it as an event of momentous significance; few people know it, even fewer have experienced it. It falls, however, under the rubric of what Stiegler describes as a "transmittable knowledge of care," a "negentropic différance" (2019: 259) that in this case exemplifies a different way of anticipating and relating to technological finitude. As such, the game lingers as an event, a concretized ethos of death that, though materially dissolved, can still haunt today's imagination as a communicable trace that complicates the core tenets of frictionless design.

Considering that Jackson encourages a deeper appreciation of breakdown and repair in the interest of building a more sustainable world, it seems especially pertinent that *GlitchHiker* was created during an event organized in purview of the theme of extinction; Ismail explains that the game was conceived by distilling "ideas about a dying world down to a dying game" (quoted in Peel 2014). The game makes evident that different forms of engaging technology—illuminated by feelings of care and morality—are possible and, if we follow Jackson and Stiegler, necessary in a world under ecological duress that nevertheless seemingly cannot shed the shackles of unbridled optimization, connectivity, and consumption. As the previous chapter sought to convey, the material infrastructures required to perpetuate frictionlessness produce vast yet unevenly distributed exploitations, erasures, and environmental degradations. Stiegler is especially insistent on this latter point, stressing that today's technological culture of consumption "consists in *hastening the Anthropocene's approach towards its limits*" and that the only way out of this predicament is to excavate new means and knowledges of care

(2019: 50, emphasis in original). *GlitchHiker* shows how such knowledges could, for instance, lead us in the direction of a pharmacology that sees users cure rather than repudiate fragile technologies. The game resonates with what video game scholar Alenda Chang, in a pharmacological register, describes as a "curative potential" of video games that derives from how narratives and aesthetics of collapse can induce a more conscious attitude toward the environment (Chang 2019: 234). In the following chapter, I will further explore the notion of technologically cultivated care as a way of tempering the accelerating temporalities of obsolescence and erasure, and of producing less fragile or virulent technological ghosts. This inquiry will expand the argument that I have begun to develop here: an aesthetic of imperfection can bring the chronolibidinal constitution of care and desire to the fore in the engagement of technology, underlining the human capacity to recognize the existential primacy of imperfection and forming a possible remedy against the destructive tendencies of frictionlessness.

4

A Death Sentence Decreed in Binary Code

The Collapse of PAL and the Spectral Afterlives of Technology

One of Hungarian writer László Krasznahorkai's most acclaimed novels, *War and War* (2006), describes the tribulations of the depressed clerk Korin. Korin, after coming across a manuscript whose beauty overtakes him entirely, vows to travel from Hungary to New York—bellwether of the new, globalized world—to upload the text to the only recently popularized Internet. While this task proves onerous and frequently leaves him hopeless, we ultimately see Korin succeed in his plan: he transcribes and transfers the manuscript to a website (which can, in fact, be visited at warandwar.com) and, convinced of its endurance, returns to Europe to commit suicide.

Krasznahorkai's work, which won him the Man Booker International Prize in 2015, is relevant to the scope of my study on multiple levels. For one, Krasznahorkai's writing, driven as his stories are by ceaseless cycles of ruination, betrays a keen intuition of the ineluctable workings of autoimmunity—the constitutive condition, charted in Chapter 1, of always-already being threatened from within. Few writers, moreover, have as adeptly narrativized how political projects that promise purity and perfection always require violence to maintain their reveries—not the necessary violence of *différance*, but physical violence measured in decimated lives and livelihoods. In *War and War*, this is exemplified by the manuscript Korin discovers, which recounts the travels of four brothers-in-arms who are repeatedly confronted by the nefarious Mastemann, a figure who pledges order and reason but unfailingly brings bloodshed and disease. However, considering my interest in technological imperfection, it is a more intimate moment of tragedy that is of note here. Should a curious reader, after having finished *War and War*, attempt to visit the website to which Korin uploaded the manuscript, the following message will greet them: "[T]his home page service has been called off due to recurring overdue payment. Attempted mail deliveries to Mr. G. Korin have been returned to sender with a note: address unknown. Consequently, all data have been erased from this home page." This implosion of the illusion of digital permanence distills pathos from an assertion that has been repeated many times throughout this study: finitude and fragility are inescapable conditions, and technological continuity is therefore never guaranteed. As Chapter 3's analysis of the video game *GlitchHiker* underlined, the question of if and how a technology survives is often entirely dependent on sustained investment and attention. Korin's unfortunate fate reveals this recuperative work to be constant and to even transcend one's own finite life.

Technologies, however, seldom die as neatly as the wholesale erasure of Korin's digitalized manuscript or the irreversible passing of *GlitchHiker* might suggest. As this chapter will expound, technologies, after having been discontinued and discarded, tend to linger like ghosts. This argument is based on an analysis of Rosa Menkman's *The Collapse of PAL* (2010), an audiovisual, glitch-based performance that departs from a notion of technological finitude and that draws on an aesthetic of imperfection to critically interrogate the destructive logic behind what, in Chapter 2, I have described as the philosophy of frictionlessness. By casting light on a deadened technological standard—the PAL (Phase Alternating Line) signal, a once prevalent color encoding system for TV that has now been made obsolete—the performance aims to compensate for the cultural tendency to condemn technologies that have been deemed imperfect to disuse and obsolescence, often without the least of funereal rites.

I first briefly outline *The Collapse of PAL*'s most notable characteristics, particularly its glitch-based aesthetic of imperfection, its ties to Menkman's own theoretical work on glitch and technology, and its invocation of Walter Benjamin's canonical figure of the Angel of History. On the basis of my reading of Menkman's performance, I develop three central arguments. First, in response to Jussi Parikka and Garnet Hertz's influential and valuable notion of *zombie media*, I propose that it is even more productive to conceive of technology in ghostly terms; to think of superseded technologies not as zombies, but as *spectralized* agents that have been cast to the margins while continuing to produce material effects, accommodating multiple temporal logics. Second, I suggest that Menkman's channeling of Benjamin signifies a certain historico-aesthetic approach to digital technology, one that reads in the outmoded and marginalized traces of technological development a shadow history of destruction. Third, I argue that by virtue of the aesthetic and conceptual logic of imperfection through which she spectrally re-renders PAL's destruction, Menkman advocates a sense of *care* for technology. Expanding the scope of the previous chapter, I conclude that facilitating new bonds of care is a crucial prerequisite for building a world where technology's detrimental material effects ripple out less unevenly across a fragile globe.

The Collapse of PAL

Rosa Menkman is a Dutch scholar and artist. She is one of the leading voices on the subject of glitch theory and has also produced numerous works of art that wed theory to practice. One of her most influential artworks is *The Collapse of PAL* (2010), an audiovisual performance presented, amongst a variety of other venues and platforms, on national Danish television and at renowned digital arts festival Transmediale. This performance was commissioned as part of an art project that explored Denmark's switch in television broadcast, in 2009, from Phase Alternating Line (PAL) to Digital Video Broadcasting (DVB) (Menkman 2016a: 116). PAL is an analog color encoding system for television that once counted as a common standard but that has since been supplanted by more efficient digital alternatives. Most countries that used to rely on it have long since switched to DVB, and the last remaining countries still to employ it are generally in the process of transitioning. Menkman's performance thus documents the fate of a technology that has been rendered obsolete, replaced by a more frictionless alternative.

The Collapse of PAL consists of three separate acts, respectively labeled "Eulogy," "Obsequies," and "Requiem for the Planes of Blue Phosphor." Multiple versions of the performance are available on the internet—I am basing my reading mostly on the various segments uploaded to the project's webpage (Menkman 2018) and on fragments available via YouTube (particularly of a 2011 performance at the now-defunct Amsterdam-based nightclub Trouw). Menkman herself describes the plot of the performance as follows:

> In "The Collapse of PAL" (Eulogy, Obsequies and Requiem for the planes of blue phosphor), the Angel of History (as described by Walter Benjamin) reflects on the PAL signal and its termination. This death sentence, although executed in silence, was a brutally violent act that left PAL disregarded and obsolete. While it might be argued that the PAL signal is dead, it still exists as a trace left upon the new, "better" digital technologies. PAL can, even though the technology is terminated, be found here as a historical form that newer technologies build upon, inherit, or have appropriated from. Besides this, the Angel also realizes that the new DVB signal that has been chosen over PAL is different, but at the same time also inherently flawed as PAL. (2012)

While the work's explicit invocation of Walter Benjamin is perhaps its most immediately notable aspect, I will briefly bracket this matter to first discuss *The Collapse of PAL*'s aesthetic qualities. The performance, in

Figure 4.1 Screenshot of *The Collapse of PAL*. Image source: www.youtube
.com/watch?v=5-XVkI1z1m8. Accessed October 31, 2022.

which Menkman often employs multiple screens, is generally preceded by
a title card that is reminiscent of the start-up image of old NES (Nintendo
Entertainment System) or Commodore 64 games, recalling a time in history
when 8-bit graphics and white noise were household banalities (Figure 4.1).
This archaic aesthetic is representative of the performance as a whole; it has
been made by tinkering with a wealth of different, outmoded technologies,
such as a "broken photo camera," a "cracklebox," an "NES," and the old
European paging system "eurosystem" (2012). These disparate antiquated
technologies that, like PAL, have been consigned to the peripheries of
the technosphere, coalesce to stage a glitch-filled performance in which
everything resists definition and clarity.[1] The performance is mostly steeped
in hues of purple and blue, and the screen seems persistently saturated,
continually at risk of breaking down (Figure 4.2). All this is aurally
compounded by a noise-based soundtrack, composed of the sounds of
technological failure and distortion. By dint of its glitch- and breakdown-
filled appearance, *The Collapse of PAL* embodies both an aural and visual
aesthetic of imperfection that, through its flawed transmissions, generates a
notable sense of friction.

---

[1] As I have discussed in the previous chapter, the notion of glitch marks a moment of
perceptible interference in the routine operation of a technology and is commonly
described both in terms of imperfection and in terms of the ghostly.

Figure 4.2 Screenshot of *The Collapse of PAL*. Image source: www.youtube .com/watch?v=5-XVkI1z1m8. Accessed October 31, 2022.

Figure 4.3 Screenshot of *The Collapse of PAL*. Image source: www.youtube .com/watch?v=5-XVkI1z1m8. Accessed October 31, 2022.

In terms of content, the performance toggles between shots of distorted landscapes that are replete with looming shapes (Figure 4.3) and shots of Menkman's reimagination of the Angel of History (Figure 4.4). As stated, this figure is borrowed from the work of Benjamin ([1955] 2015), and Menkman

Figure 4.4 Screenshot of *The Collapse of PAL*. Image source: www.youtube .com/watch?v=5-XVkI1z1m8. Accessed October 31, 2022.

presents her[2] as a vague and ghostly presence that mournfully regards the many destructions she encounters. This sense of mournfulness is conveyed not simply through the ruinous atmosphere of the performance but also through the narrating text overlaid onto it (Figure 4.5), which delivers the following monologue:

> *The angel of History had television*
> *She witnessed the termination of PAL*
> *And when the PAL signal was muted*
> *Its chance to clarity smothered*
> *A brutal but silent execution had taken place*
>
> *The Angel would like to stay*
> *And awaken the dead connection*
> *Make whole what has been broken*
> *But a brutal storm is blowing towards utopia*
> *It caught her wings with strength*
> *And the angel can no longer close them*

---

[2] While Benjamin refers to the Angel as a man, Menkman frames the Angel as a woman. Media scholar Tiffany Funk, in her reading of *The Collapse of PAL*, focuses on this presentation of the Angel and of PAL as "female presence[s]" and argues that they indicate a critique of how "patriarchal values are . . . embedded within the histories of our current media technologies" (2018: 165).

Figure 4.5 Screenshot of *The Collapse of PAL*. Image source: www.youtube
.com/watch?v=5-XVkI1z1m8. Accessed October 31, 2022.

The storm propels the Angel backwards
Brushing history against the grain
Into a new horizon of collapse
In front of her
She sees a pile of debris
Growing skyward
Connections that were just not good enough
They are now left behind
To loose [sic] their significance

PAL slowly vanishes in these eerie ruins
Only to survive as a trace
A memory left onto other connections
Crashed and collided
This is where PALs history
Can still be found
A lost signal

———

Spoils of the Hype Cycle
The exploitation of errors has become a standardized norm

The dogmatic search for the perfect signal
Is ill-fated

All technologies have fingerprints
Of imperfection
Noise artifacts
Glitches, feedback and compressions
These artifacts can be exploited
As a new language

Menkman presents a narrative of technological decay and perseverance; of a technology sentenced to death that nonetheless still lingers and is traceable should one seek it out. Her explicit intention with *The Collapse of PAL* was to introduce a melodramatic arc to a piece of glitch-based art in order to show how glitch's aesthetic of imperfection and failure can carry cultural and theoretical weight (Menkman 2011b: 8). Some of the narrative's sentences have been lifted (almost) verbatim from Menkman's theoretical text *The Glitch Moment(um)*, which redoubles the intimate relation between theory and practice within her work. By tying an aesthetic of imperfection to a more conceptual sense of imperfection and finitude—one that is informed by the phenomenon of technological obsolescence—the performance resonates with the object of *GlitchHiker* that was discussed in the previous chapter. Yet, whereas *GlitchHiker* forged an emotional connection to its own ephemeral existence, *The Collapse of PAL* appears to make a more general argument about how the modern world deals with its technologies, and particularly with those technologies it deems imperfect. At heart, the performance fosters an awareness that, even when disavowed, a technology never truly dies and may therefore manifest itself or be sought out in a myriad of ways even after the ostensible fact of its termination.

This last point is made palpable by a brief essay that Menkman wrote for the 2016 *Transmediale Reader*, titled "Elegy for the Collapse of PAL (2010-2012)." In this essay, Menkman describes how, after years of enforced silence, PAL and the Angel of History can now finally communicate again, owing to the arrival of a new technological application: "Since the switch to DVB, the Angel of History has not been able to transmit to PAL. Recently, however, a technology named Syphon made it possible to broadcast and connect different formats and signals—old, new, and even obsolete—via local servers. For the first time in years, PAL and the Angel of History can now reconnect" (2016a: 116). Menkman offers a transcript of the ensuing conversation, wherein PAL imparts that it sensed the Angel was trying to reach it but that it could not receive its broadcast, and the Angel explains that it lives on in a world still flooded with new technologies and updates that nonetheless retain their immanent imperfections (2016a: 120). The conversation is tragically cut short when, just as PAL and the Angel pledge to symbiotically re-render

The Collapse of PAL, Syphon starts to malfunction and communications are lost. Menkman's short text accentuates PAL's liminal state; even though cast to the margins, PAL never fully disappeared and new conditions suddenly allow it a means to assert itself. The text, however, equally suggests this new connection to still be fraught with contingencies, dependent on the precarious materiality of mediation. "Elegy" was later translated into a performance and staged in New York City with an accompanying video and a material shrine, "featuring blue, burned cassette tapes" (Menkman 2018), that further emphasized PAL's place in a wider history of culturally rejected and discarded objects that can still be reappropriated.

The Collapse of PAL is a performance that is rife with ghosts and orchestrates a form of hauntological aesthetics. In keeping with the way I defined this aesthetic approach in Chapter 2, Menkman's work draws on an aesthetic of imperfection to consider how technology impacts perception, memory, and anticipation, and explores the attendant cultural attitudes and possible alternatives. In the previous chapter, I augmented the notion of hauntological aesthetics by specifying three spectral qualities that account for the popular association of glitches and ghosts: glitch's ability to highlight technological norms that go unheeded; glitch's capacity to disclose spectral forms of technological agency; and glitch's tendency to presage death. In the case of *The Collapse of PAL*, the first and third quality are especially relevant. Appealing to the former quality, Menkman herself maintains that glitch is concerned with mapping out that which evades or exists beyond established knowledge (2011: 66). In a kindred spirit, Avery Gordon emphasizes that to probe for ghosts "is a case of the difference it makes to start with the marginal, with what we normally exclude or banish, or, more commonly, with what we never even notice" (1997: 24–5). This link between ghost and glitch captures the pedagogical and hauntological stance that Menkman's spectral performance assumes: its glitches underscore the violent act of PAL's termination—an act "executed in silence"—and amplify it. The performance advocates, in other words, an epistemological perspective that does not abide by the covert erasures that frictionlessness so often ordains, but, instead, seeks knowledge of the tenebrous (after)lives of a technology condemned to oblivion. This notion of technology as something that can be terminated also betrays the other relevant ghostly quality of glitch: its ability to figure a form of technological finitude. As elaborated in Chapter 3, glitches signal the always possible prospect of breakdown and obsolescence. No technology is free from the condition of autoimmunity, meaning technological endurance and fidelity can never be total, a predicament that is echoed by *The Collapse of PAL*'s suggestion that "all technologies have fingerprints of imperfection." The most important implication of this glitch-based signification of finitude, so I will now contend, is that the performance suggests that technological finitude seldom spells an act of absolute disappearance and that PAL *itself* survives as a kind of ghost.

Of Ghosts and Zombies

Media historian Jussi Parikka frames Menkman's art as a prime example of the discipline of media archaeology (2012: 151). Media archaeology, a form of "media analysis *of and from the ruins*" (Parikka 2012: 90, emphasis in original), describes both a field of studies and an artistic practice that surveys the material conditions of medial production and the way technological pasts still act on the present. Parikka maintains that media archaeology "can be used to find the neglected . . . in the midst of celebrations of communications and *frictionless* digital culture" (2012: 91, emphasis added), although he does not further develop this notion of the frictionless. "Menkman's work with archaeologies of technical signals, image protocols and glitch aesthetics," Parikka proceeds to argue, has a doubly critical capacity, "excavating longer time-spans in order to understand the conditions for the contemporary scientific media culture, and . . . excavating the technicalities of current technologies in order to understand how they frame our world" (Parikka 2012: 151; cf. Cubitt 2014: 47–8). This claim already reveals an affinity between media archaeology and hauntological aesthetics, as both fields foster practices that delve into technological temporalities and contingencies. Menkman's work is part of a wider movement of media-archaeological artists who survey the material legacies of media and disturb linear conceptions of time—notable examples include Sarah Angliss, Kristoffer Gansing, and Holly Herndon.[3]

Menkman invokes the discipline of media archaeology when she states that PAL, once it became obsolete, "became part of the 'zombie media'" (2016: 116). This term refers to an influential concept developed by Parikka and artist Garnet Hertz. Zombie media, they explain, denotes "the living dead of media history and the living dead of discarded waste that is not only of inspirational value to artists but signals death in the concrete sense of the real death of nature through its toxic chemicals and heavy metals" (Hertz and Parikka 2015: 145–6). Zombie media, in other words, in large part concerns the "afterlife of retired high-tech objects" (Kane 2019: 176). There is a twofold logic to the living-dead traces that Hertz and Parikka describe. First, the

---

[3] For exemplary works on media archaeology, see Elsaesser (2016); Kittler (2013).

idea of living dead illustrates the way in which a technology reaps material effects even, and often especially, after its (usually planned)[4] obsolescence arrives, although these effects tend to be kept from the consumer's view. As Chapter 2 illustrated, technological finitude is often only a localized finitude, as disused consumer products are freighted to occulted sites of disposal where they remain to affect an already vulnerable environment. Zombie media, indeed, spells out that technologies never die, at least not in the sense of a clean and total disappearance (Hertz and Parikka 2015: 153). Second, as demonstrated by *The Collapse of PAL*, the living-deadness of zombie media suggests that technologies live on as traces in newer technologies and that, in line with the sensibilities of hauntological aesthetics, there always remain aesthetic and artistic ways of reappropriating technology, of probing it for alternate presents and futures.

The undead figure of the zombie that lends this mediatic logic its name, indeed, presents an imposing conflation of life and death, and its associations of virality especially befit a (post)pandemic world. Yet, Menkman's performance and its hauntological aesthetic raise the question of whether the zombie truly provides the most productive metaphor through which to understand the multiplex temporalities and material sedimentations of purportedly deceased technologies. Is not the ghost, as it has thus far been described, a more potent descriptor for how technologies reign from beyond the grave, both as environmental agents and as artistically reappropriable traces? Inspired by philosopher Donna Haraway's assertion that "it matters what stories we tell to tell other stories with" (2016: 12), I believe that there are at least four reasons why the ghost offers a more fruitful conceptual and narrative template than the zombie does for thinking through the afterlives of media.

First, while the zombie certainly conveys a confused temporality of life and death, the ghost affords a more heterogeneous palette of often conflicting temporalities. Ghosts, as I explained in previous chapters, bespeak complex

---

[4] The idea of planned obsolescence, of purposely delimiting the life span of a product to stimulate consumption, is a central subject of media archaeology's critical inquiries. As Hertz and Parikka recount,

> [i]n the United States, about 400 million units of consumer electronics are discarded every year. Electronic waste, like obsolete cellular telephones, computers, monitors, and televisions, composes the fastest growing and most toxic portion of waste in American society. As a result of rapid technological change, low initial cost, and planned obsolescence, the federal Environmental Protection Agency (EPA) estimates that two-thirds of all discarded consumer electronics still work—approximately 250 million functioning computers, televisions, VCRs, and cell phones are discarded each year in the United States. (2015: 141–42)

webs of past, present, and future, and are intimately related to the trace-like logic of time itself. Following the Derridean logic of spacing, the present is always-already contaminated by both past and future, locating a logic of haunting in the heart of the temporal. If, as cinema theorist Vivian Sobchack stresses, the inquiries of media archaeology reveal "the past as in some way always present" and "the present and future as in some way already past" (2011: 325), they accordingly reveal technology to be traversed by spectrality. Menkman's performance indeed emphasizes that PAL slowly vanishes but remains as a faint trace imprinted onto newer forms, implying that technologies speak in the spectral tongue of what precedes and succeeds them. Zombies, on the contrary, signify a more discrete temporal logic that connotes a crude demarcation between before and after. These entities, furthermore, often spell a situation of being stranded in time, the future forever failing to appear—as film scholar Evan Calder Williams argues, "[t]he anxiety proper to zombie films is the deep horror of something not being different, of *everyone* remaining as limited a category as we know it to be, of the same persisting, of the end of death and lack" (2011: 103, emphasis in original). While ghosts and hauntology may similarly signal a failure of the future, they also grant the hope and possibility of a rupture, a reconfiguration of traces that could lend lost or imagined futures a more material anchorage. A neglected technology, moreover, remains impactful on many different timescales, from the direct fact of its disposal to the deep times of geological erosion to which it contributes. This is in fact supported by Parikka's own leverage of the term "remain(s)" as both a verb and a noun that illustrates how disused technology retains the "liveness of multiple afterlives" (Jucan, Parikka, and Schneider 2019: 43). While the afterlife of the zombie is generally restricted to one decomposed body, a dead technology can, like a ghost, branch out in many different temporal directions.

Second, ghosts and technologies can spread out in multiple *spatial* directions as well. While zombies are singular, identifiable, and often achingly conspicuous beings, ghosts can inhabit a range of different forms. This multiplicity better approximates the realities of technological finitude: technological devices, in their ostensible hereafter, are stripped for parts; they are disassembled and rearranged; they rust and rot; they lie forgotten in dust-filled attics, or reenter the commodity cycle as nostalgic curiosities; and, once inhabiting the landfill that seems their final resting place, their components slowly detach themselves from one another and, dissolved into air and poured into soil, commence a new toxic afterlife. These manifold processes appeal more to the ghost, as a figure that inhabits many different shapes and that cannot be tied down, than to the zombie, as a figure that remains integrally traceable and perceptible. Relatedly, contrary to the zombie's restless shuffling, ghosts permeate, leak, vaporize, imprint, and proliferate, and thus yield a more dynamic conceptual lens. The ghost, in short, adheres more closely to Hertz and Parikka's own assertion that "media never die; they decay, rot, reform, remix, and get historicized, reinterpreted, and collected" (2015: 153).

Third, the zombie as we know it from popular cultural is almost solely something to be reviled and exterminated, and almost never a creature to willingly share the earth with. As the zombie's lore tells us, a well-timed shot through the head generally appears to be the only established response. Conversely, while cinema and literature certainly feature many malevolent ghosts that justify an exorcism, there are also myriad examples of ghosts that stimulate a different reaction (reparation, amendment, amusement, retreat). For instance, as Esther Peeren attests, there are numerous groups whose state may be represented as ghostly (migrant laborers, missing persons, domestic servants) that warrant an ethics of attention and care rather than castigation (2014: 8–9). To conceive of technology as spectralizable, as able to take on but also to be inflicted with ghostly traits, makes it easier to think of depreciated technologies as, literally, a matter that we must remain with. Such a perspective calls for a responsible demeanor that takes into account how technologies continue to produce material effects even, or especially, when turned into waste and demands an effort not to let such technologies slip from view to produce death elsewhere. This technological claim for justice is something that is underlined by Menkman's performance; she presents PAL's demise as a "brutally violent act" that the Angel of History condemns, effectively framing PAL as a ghostly stain that attests to past wrongdoings.

Fourth, the zombie is traditionally a figure that lacks agency, presented as a brainless being that has no command over its own actions, a physical shell with no ghosts left in its machine. The zombie appears emptied of all virtual forces; only the body and its impulses remain. Ghosts, on the contrary, generally suggest both the erasure of their once material, visible body and a proliferation of environmental effects. Against the limited agency of the zombie, ghosts may partake of the variegated affordances of what Peeren terms "spectral agency" (2014: 16). The concepts of ghost and haunting grant a significantly richer template to make sense of the way in which outmoded media impact the earth, especially for haunting's connotations of material effects produced by something that remains insensible or unacknowledged. Again, this resonates with Parikka's own account, as he underlines that the remainder of media "does not merely mark a spot of something gone but works as its own generative force" (Jucan, Parikka, and Schneider 2019: 9). Read in line with the previous paragraph, this further underscores the significance of attending to the question of if and how a specific technology should be spectralized; if neglected or disposed of, what possibly adverse effects may this technology reap and what agencies will be kept from our perception?

For these reasons, I suggest that ghosts and the idea of spectralization offer a more productive description of conditions of technological obsolescence than the zombie does. In previous chapters, I described spectralization primarily as a process of rendering invisible, liminal, and/or fragile. Such a spectral perspective yields numerous vital benefits for conceptualizing and understanding the effects of technological obsolescence: it provides a register through which to point to processes of marginalization, to assess how that which has ostensibly disappeared still makes material claims and to seek new, more just relations to what has been violently displaced. While never explicitly describing PAL as such, *The Collapse of PAL* makes palpable that a technology can be turned into a ghost: discarded and denounced, but continuing to act, and traceable if one proves willing and responsive. This notion is strengthened by the performance's hauntological aesthetic, which besets its audience with imposing shapes that ambivalently assert themselves and with hidden agents that constantly threaten to break through the noise. The performance, furthermore, presents PAL's ghostly fate as a grave injustice, a silent "death sentence" that comprised an act of great violence. This latter point brings me to the second argument I will develop in relation to *The Collapse of PAL*. Having made the performance's implicit suggestion that technology can be rendered spectral explicit, I will now address its invocation of another ghost: that of Walter Benjamin's Angel of History. This figure, I will demonstrate, allows Menkman to unfold an artistic and historico-aesthetic perspective premised on imperfection.

A History of Frictionless Destruction

What does *The Collapse of PAL*'s latent suggestion that technology can be spectralized convey? And what kind of hidden or neglected knowledge does the performance's hauntological and glitch-based aesthetic seek to divulge? Above all, I posit that the performance attempts to make explicit the destructive undercurrents of technological progress. More specifically, Menkman's artwork aestheticizes the destructive effects of the ongoing intensification and expansion of the design philosophy of frictionlessness.

The Collapse of PAL's concern with technological destruction is epitomized by its reimagination of the seminal figure of Walter Benjamin's Angel of

History.[5] Benjamin originally invoked the image of the Angel, inspired by Paul Klee's 1920 painting *Angelus Novus*, in reference to the social and political horrors of his time:

> A Klee painting named Angelus Novus shows an angel looking as though he is about to move away from something he is fixedly contemplating. His eyes are staring, his mouth is open, his wings are spread. This is how one pictures the angel of history. His face is turned toward the past. Where we perceive a chain of events, he sees one single catastrophe which keeps piling wreckage upon wreckage and hurls it in front of his feet. The angel would like to stay, awaken the dead, and make whole what has been smashed. But a storm is blowing from Paradise; it has got caught in his wings with such violence that the angel can no longer close them. The storm irresistibly propels him into the future to which his back is turned, while the pile of debris before him grows skyward. This storm is what we call progress. ([1955] 2015: 249)

This famous paragraph captures Benjamin's philosophical and historical vantage, always marked by a deep sense that it is the outmoded, the obsolete, the imperfect from which the destructive course of history can be divined. Refusing to be seduced by the promises of progress, Benjamin preferred instead to stay with "the rags, the refuse" ([1982] 2002: 460, N1a,8) of mass culture: for him, the shameful undercurrents supporting any illusion of advancement were revealed by precisely those objects whose aesthetic and technological qualities had been overtaken by newer, more impeccable objects.

Before further exploring these claims, it is important to note that—for all the religious overtones that surround his figure of the Angel (and the messianic spirit of his wider work)— Benjamin's perception of history and investment in the obsolete negated responses to finitude that sought transcendent salvation rather than material and political investment. This is already evident from his early work on the tradition of the German *Trauerspiel*, where he analyzed how Baroque poets allegorized images of ruin to reveal transiency as the unconditional usher of history. Ruins are sites where time has left its indelible mark, making plain that nothing is safe from its effects: "In the ruin history has physically merged into the setting. And

---

[5] Benjamin's Angel of History also appears in the previously discussed work of science and technology scholar Steven Jackson. The figure of the Angel, so Jackson proposes, exemplifies a "broken world methodology" (2014: 237) that tends to the fractured rather than fetishizes the immaculate.

in this guise history does not assume the form of the process of an eternal life so much as that of irresistible decay" (Benjamin [1928] 2009: 177–8). However, for the Baroque artists he surveyed, the ultimate decay of matter prompted a turn to the spiritual realm: "Ultimately in the death-signs of the baroque the direction of allegorical reflection is reversed; on the second part of its wide arc it returns, to redeem" (Benjamin [1928] 2009: 232). With this "about-turn, . . . the immersion of allegory has to clear away the final phantasmagoria of the objective and, left entirely to its own devices, re-discovers itself, not playfully in the earthly world of things, but seriously under the eyes of heaven" (Benjamin [1928] 2009: 232). Benjamin could not align himself with this rejection of the fragile physical world in favor of a divine transcendence of materiality. On the contrary—and this is what his Angel of History illustrates—Benjamin saw in the world's irredeemable condition of imperfection and mortality an impetus to attend closely to historically specific impositions of vulnerability and finitude. Each historical epoch should, in Benjamin's words, be evaluated by "interspersing it with [its] ruins" ([1982] 2002: 474, N9a,6), unearthing the oft-concealed forms of destruction that have facilitated it.[6]

Menkman's invocation of Benjamin and presentation of PAL as a technology turned spectral seek to redouble this gesture in the context of the digital. Not only does Menkman summon the Angel aesthetically to play a prominent role in the performance, but she also paraphrases Benjamin's words throughout the narrating text. As the narrative of the performance recounts: "the angel would like to stay and awaken the dead connection"; "but a brutal storm [. . .] caught her wings with strength"; "she sees a pile of debris growing skyward." These sentences resound with those of Benjamin ("the angel would like to stay, awaken the dead"; "but a storm is blowing from Paradise; it has got caught in his wings"; "while the pile of debris before him grows skyward") and reveal a perspective that highlights the destruction immanent to technological progress (Benjamin [1955] 2015). The "utopia" to which the performance's Angel is forcefully blown is of a dubious nature as it leaves a multitude of victims in its trail. With a mournful gaze, the Angel

---

[6] If one thing characterized Benjamin's philosophical orientation, it was his inclination to approach history and its wreckages with the intention to commune with ghosts. While I cannot fully account for the many relations between Benjamin's work and the spectral here, it is worth signaling that Benjamin betrayed a profound intuition of the spectral constitution of the temporal—his work "blasts through the rational, linearly temporal, and discrete spatiality of our conventional notions of cause and effect, past and present, conscious and unconscious" (Gordon 1997: 66)—and of the many ghosts this constitution produces, believing that confronting the present with all that it marginalizes and rejects brings it "into a critical state" (Benjamin [1982] 2002: 471, N7a,5).

regards the "brutal but silent execution" of PAL as one more calamity in a world already composed of ruins. The "pile of debris" that Benjamin made his Angel contemplate, distended by "wreckage upon wreckage" ([1955] 2015: 249), thus finds its mirror in the performance; Menkman's reinterpretation of the Angel, the performance relates, sees "a pile of debris," a technological realm of "eerie ruins" that PAL is forced to inhabit.

While most obviously indebted to Benjamin, Menkman's tale of digital destruction also resonates with a more recent conceptualization of history—one that is spiritually akin to Benjamin's work but that also discloses how, in digital times, the occultation of destruction that troubled Benjamin has further intensified. One of literary scholar Alan Liu's main arguments in *The Laws of Cool* is that today's technological culture is increasingly unwilling or unable to offer serious reflection on the many demolitions that occur in its wake (2004: 322). There are few threnodies to be found for whatever ends up being diminished under the relentless demand for digital perfectibility. As Liu suggests, the "postindustrial ethos of creative destruction" has become general and the economic demand for quick turnover necessitates a continual erosion of past and present (2004: 375). It is in this context that Liu signals a unique opportunity for an "alliance of the arts and contemporary humanities" (2004: 375), convening to bring historical revivification to what otherwise remains subdued:

> Where once the job of literature and the arts was creativity, now, in an age of total innovation, I think it must be history. That is to say, it must be a special, dark kind of history. The creative arts as cultural criticism (and vice versa) must be the history not of things created—the great, autarkic artifacts treasured by a conservative or curatorial history—but of things destroyed in the name of creation. (2004: 8)

This passage proposes exactly the type of historico-aesthetic practice that *The Collapse of PAL* expresses; against a technological culture of constant innovation that seeks to conceal its ruinous ventures, Menkman posits the meticulous and careful work of exhuming ghostly traces of destruction, granting them a fuller form in order to address the present. This is supported by her heavy use of glitches; she employs their epistemic quality to highlight failure and imperfection, and suggests they can give rise to a "new language" that critically interrogates the "histories of 'progress'" and the elements these histories obscure (Menkman 2011b: 44). Glitch, as she describes, encourages an attitude of "generational destructivity" (Menkman 2011: 35), of seeking knowledge by highlighting friction, contingency, and decay. This idea of generational destructivity, emphasizing rather than obscuring destruction, forms a counterweight to the muted eradications that Liu condemns. Within *The Collapse of PAL*, the frictional, spectral quality of glitch to illuminate entrenched norms and to aestheticize finitude thus combine to gesture at a certain techno-logic of silent ruin that remains dominant today.

Should, however, such a historico-aesthetic practice give rise to an egalitarian attitude whereby all objects are to be equally restored and shielded from obsolescence? Pertinently, for Benjamin, it was not that all objects should indiscriminately be saved from desuetude, nor was this desuetude in itself his primary focus:

> What are phenomena rescued from? Not only, and not in the main, from the discredit and neglect into which they have fallen, but from the catastrophe repressed very often by a certain strain in their dissemination, their "enshrinement as heritage."—They are saved through the exhibition of the fissure within them.—There is a tradition that is catastrophe. ([1982] 2002: 473, N9,4)

What was most significant about the objects Benjamin salvaged was, in other words, not the mere fact of their disuse, but rather the catastrophes that were concealed by their inclusion in a sanitized narrative of technological progress. Catastrophe, for him, consisted in the historic production of violence but also in the ongoing domination, through suppressing technology's democratic and open-ended potentials, of societies by a technocratic elite. Menkman, similarly, seems less concerned with the PAL signal as a technology that would somehow be uniquely deserving of redemption, and frames PAL's collapse more as symptomatic of a larger logic of technological destruction. She appears to lament a culture of wastefulness and a stifling of the imagination, where people fail to consider the full potentials and implications of technology; users follow the prescribed pathways of "built-in obsolescence and built-in nostalgia," where technologies are rapidly rejected and occasionally heralded back as nostalgic commodities, only to be discarded again when the "hype cycle" proclaims a new fetish (Menkman 2011: 57; cf. Reynolds 2011). Accordingly, Benjamin's century-old claims take on new relevance within *The Collapse of PAL*; the performance decries a technological disposition whereby past devices are quickly shipped from view or are neatly lodged in the cultural imagination as nostalgic signposts on the road to technological perfection, with little eye for whatever corrosive processes their demise signifies or for the unrealized potentials these objects contain.

Specifically, I posit that Menkman's performance apprehends and challenges the destructive and ghostly nature of the philosophy of frictionlessness as mapped in Chapter 2. Frictionlessness, to reiterate, is a design philosophy that aims to tie perception and activity seamlessly to the operations of technology while hiding its destructive tendencies from view. It achieves this primarily by culturally instilling or intensifying the values of user-friendliness, connectivity, and optimization. I have explored how frictionlessness is essentially a spectralizing philosophy, insofar as its smooth surfaces operate only by virtue of the toil of a subliminal realm of specters, a predicament that is aggravated by its tendency to condemn technologies to obsolescence and to compel the production of the new. *The Collapse of PAL* highlights one such rejected technology and considers the cultural logic that has caused its spectral state.

The Collapse of PAL explicitly ties PAL's demise to the value of connectivity; not only does its website describe it as being "about the loss of connectivity" (Menkman 2018), but the narrative of the performance also emphasizes that the derelict expanse through which PAL dwells is strewn with objects that supported "connections that were just not good enough." It is, then, the frictionless logic of glorifying connectivity—of seeking ever-smoother linkages and transmissions—that has precipitated these myriad extinctions. Paradoxically, PAL became a specter because it could not function spectrally enough; it was overtaken by digital alternatives that promised more transparent, efficient, and imperceptible modes of mediation. Implicit behind this process is the value of optimization, driven by the "dogmatic search for the perfect signal" (Menkman 2011: 11), whereby technologies whose connective capacities prove inadequate by present-day standards are to be quickly cast away. The performance underlines that this process is unending; intuiting the autoimmunity of technology, the narrative describes "the dogmatic search for the perfect signal" as "ill-fated" because "all technologies have fingerprints of imperfection." *The Collapse of PAL* thus exposes frictionlessness as a particular rationale of mobilizing the inescapable conditions of imperfection and finitude: because no technology could ever provide a fully frictionless, connective experience, and because obsolescence can always be made into an integrated function, the destructive cycle of optimization and consumption is perpetual.

Menkman attempts to draw more reflective attention to the ghostly destructivity of this logic. In "A Glossary of Haunting," cultural theorists Eve Tuck and C. Ree underline the importance of making manifest how historical destructions and the ghosts they breed often escape perception:

> [W]e are always in a process of ruin, a state of ruining. Our ruins are
> not crumbled Roman columns, or ivy covered [*sic*] abandoned lots. Our
> ruins lie within the quick turnover of buildings, disappearing landmarks,
> and disposable homes, layered upon each other and over again. And in

the tradition of the symbolism of horror, the ruin always points to the scene of ghost-producing violence. The ruin is not only the physical imprint of the supernatural onto architecture, but also the possessed or deluded people wandering amidst the ruin who fail to see its ruinous aspect. (2016: 653)

While this passage moves rather ambiguously from ruin as an interminable process ("we are always in a process of ruin") to ruin as a historically situated locale, it reveals the pertinence of Menkman's implicit presentation of PAL as a kind of ghost and of her figuration of technological development as a process that ceaselessly consumes its own devices. Tuck and Ree maintain that, even as one lives under conditions of destruction, it requires a certain perceptiveness and responsiveness to note the present's "ruinous" and "ghost-producing" nature (2016: 653). The philosophy of frictionlessness, exporting its ruinous traces away from the consumers that are its central capacitators, actively deters such perceptiveness. Its pristine and immersive devices are aesthetically designed to deny that any violence undergirds them or that any alternate potential is locked away inside them. Menkman's performance, conversely, seeks to floodlight the technological ghosts engendered through the fetishization of frictionless transmission. If we prove willing, we can recognize these ghosts, and the ruins through which they dwell, in "the rapid turnover" (Tuck and Ree 2016: 653) of devices and applications, in the unceasing cycles of obsolescence and renewal that technology demands. *The Collapse of PAL*, in sum, highlights how, within the far-reaching "scene of ghost-producing violence" (Tuck and Ree 2016: 653) that Chapter 2 showed frictionlessness to be, many technological devices are themselves eventually spectralized in the rush of optimization.

While Menkman herself does not explicate this, the wider field of media archaeology shows that one of the most troubling effects of such spectralizations is that, when removed from view, technological ghosts may proceed to cause material damage—think, for instance, of the e-waste produced by the many discarded televisions that employed PAL as a standard (Hertz and Parikka 2015: 142). In fact, it is in this context that I consider the aesthetic of imperfection and the frictions of Menkman's work to be most critical. *The Collapse of PAL*'s aesthetic qualities resonate with communications scholar Carolyn L. Kane's recent suggestion, made in reference to the environmental destructions outmoded devices continue to reap, that "an aesthetic of failure may be our most viable option for accepting the realities of the present and a prerequisite for sustained change" (2019: 24). Such an aesthetic perspective does not prescribe a naive admiration for the run-down but encourages the hard work of attending to what is broken or rejected, assuming responsibility for the consequences of a technological culture of neglect. More specifically, I propose that, expanding the arguments developed in the previous chapter, Menkman's work effectively advocates a form of *care* for technology that the philosophy of frictionlessness structurally forestalls.

Care in Times of Frictionlessness

By presenting PAL as a marginalized ghost subjected to injustice and by invoking Benjamin's Angel of History—a figure that testifies to veiled destructions and that longs, as the performance states, to "awaken the dead connection," to "make whole what has been broken"—*The Collapse of PAL* criticizes the way many societies relate to their technologies. The Angel embodies a desire to nourish rather than neglect technologies that have been left by the wayside, but the performance makes clear that this desire is countermanded by the relentless storm of progress. The winds of optimization and innovation blow ever forward, inhibiting the careful, conscionable, and protracted work of caring for what is left behind. Menkman's work aesthetically concretizes Kane's admonition that, in a world under environmental duress that nevertheless continues to celebrate high-tech expansion, there is an urgent need for "reminder[s] to keep in check highfalutin ambitions to innovate ever-greater, newer, and faster technologies without properly caring for the afterlife of our current ones" (2019: 24).

In Chapter 1, I underlined why it is crucial that humans remain conscious of the forms their technologies assume and the materials they exploit. As beings that are uniquely capable of recognizing and orienting themselves toward the existential primacy of imperfection, humans prove themselves sensitive to the reality of loss and death in a morally inflectable fashion. While technology has no sense of the ghosts it creates, humans understand what it means to perish, suffer, and lose, and are therefore able to form chronolibidinal attachments. Chronolibido, a concept developed by philosopher Martin Hägglund that I introduced in Chapter 1, denotes that care and desire are inextricably bound to the apprehension of finitude; it is only because we sense that something can be lost or altered that we can become invested in its being. Aestheticizations of temporality and finitude— often marked by imperfection—can, for this reason, adjust humans to their environment in affirmatory ways; they can make them more invested in the world's fragile state and can urge them to prevent sufferings and erasures. These are faculties that technology lacks, and the progressively transparent endeavors of frictionlessness should, accordingly, compel us to consider how to remain in a responsible relation with technology's material unfurling.

In the previous chapter, I analyzed the video game *GlitchHiker* through the lens of chronolibido. I explained that the reason why this digital object prompted emotional investments was that it conceptually and aesthetically incorporated a palpable logic of finitude. *GlitchHiker* demonstrated how absorbing finitude and its material effects more meaningfully into technological designs can give way to a more sustainable investment that sees users care for rather than discarding what is at risk of breaking down. Like *GlitchHiker*, *The Collapse of PAL* is an object that appeals to us as chronolibidinal beings; it is an artwork that renders finitude palpable and that encourages a reflection on technological loss and on what this loss might entail. By situating technological finitude and fragility at the forefront of perception, *The Collapse of PAL* envisions technology as something that requires sustained attention if we want to remain in a responsible relation with it. This vision underlines the pertinence of the human capacity for care in relation to technology.

Care, as feminist science and technology scholar María Puig de la Bellacasa has convincingly argued, "joins together an affective state, a material vital doing, and an ethico-political obligation" (2017: 42). Accordingly, caring for something implies both an emotional investment and a material practice. Philosopher of ethics Sandra Laugier contends that care is at the heart of ordinary life; in a world that sings the neoliberal praises of autonomy, independence, and self-sufficiency, care comprises the often invisible and disregarded critical infrastructure that forms the precondition for all flourishing (2015). The centrality of care to the conduct of life has, for instance, been made acutely clear by the Covid-19 crisis: "Care is never more visible than in those situations where it is the form of life, the 'normal' life . . . that is shaken" (2020). Care reveals a collective condition of human (and, I would add, other-than-human) vulnerability that spells out that nothing can survive long without the care of others, insulated or immunized though individuals may deem themselves (Laugier 2015). Moreover, thinking with care, like thinking with spectrality, potentially "reminds us of exclusions and suffering" (Puig de la Bellacasa 2017: 19), of how neglected or marginalized subjects are deprived of the conditions to thrive. To be sure, care does not automatically describe a neat, benevolent, or romantic process and is certainly not always fulfilling. Rather, it indicates a being invested in the material realm, with particular eye for how the collective condition of autoimmunity materially betrays stark differences in individual fragility, precarity, and death. To care is not a simple "feel-good" injunction, as to care for one thing always comes at the expense of caring for something else, and care is not always wanted; it can be smothering, paternalistic, and can produce undesired and sometimes even harmful outcomes (cf. Ticktin 2011). Furthermore, care ideally takes shape not as a one-sided imposition, but as emergent from a responsiveness to the articulations of the other. Even if it is not always appropriate or desired, and even if its effects are never guaranteed, care is, above all, about remaining responsible for how the world unfolds, for the many material costs incurred by our modes of production, and, in a specifically technological context, for how technologies differentially shape and modulate our lives (Puig de la Bellacasa 2017: 43).

This latter point illustrates how care, even though it might instinctively appear strange that Menkman advocates a sense of care for technology rather than for a living being, also inflects our relation to the technological realm. As Puig de la Bellacasa asserts, "we must take care of things in order to remain responsible for their becomings" (2017: 90). I have earlier discussed the idea of pharmacological thinking—of conceiving of technology as containing both toxic and tonic potentials—to indicate that our relation to technology is never given once and for all. Different modalities of caring (and not caring) also foster "alternative affective involvements with the becomings of science and technologies" (Puig de la Bellacasa 2017: 19). In this regard, the interest in rejected technology that animates *The Collapse of PAL* should not be seen to spell inattention to the violence imposed on bodies and ecospheres. On the contrary, consideration for the detritus of technological progress magnetizes a wealth of concern for lived experiences. While speaking of technology as something that can be unduly terminated or exposed to brutality could seem superfluous in a time of very real human and other-than-human turmoil, Puig de la Bellacasa's conception of care reveals that caring for a technology rather than casting it from view is also a way of remaining responsible for the conditions of living (and dying) it helps create. Considering the fundamental role technology plays—both in the material organization of life and death, and in cognition and perception—a viewpoint that presents technology as a spectralizable agent whose ghostly excesses can be contained through care allows one to better grasp technology's ongoing impact on human and other-than-human communities.

Pharmacological thinking and the notion of technological care recall the work of Stiegler, who has repeatedly insisted that the current technological assemblage induces a "systemic carelessness" (2019: 94), or a widespread incapacity for both individuals and political institutions to be substantially affected by the destructive effects of technology and the broader desecration of the planet. While I am skeptical of Stiegler's characterization of our time as "the era of the absence of care" (2019: 261)—surely, one has only to look at the many recent and widespread instances of activist mobilization to see that many people care deeply about justice and sustainability—his account is valuable for how it enables us to see that care *for* technology is in many ways dependent on how our capacities for care are regulated *by* technology. Because technology modulates temporal experience and, by implication, aesthetically affects our sense of finitude and fragility, technology plays an active part in shaping the horizon of care. We must be able to consciously reflect on how technology shapes the world and to actively engage with and intervene in its operations if we want to remain responsible for the widespread material effects that technology reaps. In this time of technologically ordained despoliation, Stiegler postulates that "thinking by caring" (2019: 264) forms the only remedy—that taking "*care* of the *pharmakon*" (2019: 224, emphasis in original) is the only means of nourishing technology's curative potentials while keeping its poisonous capacities in check. I have insisted that pharmacological thinking must incorporate a communal dimension; if it is to produce sustainable modes of care, it must take a holistic view of the *pharmakon* that considers not only how one's own mind and life are affected by technology but also what other lives are at stake in its current manifestation.

Developing such pharmacological modes of thinking first of all requires an inquiry into the technological "conditions of possibility of care" (Martin, Myers and Viseu 2015: 10). This inquiry must integrate the dimension of aesthetics. *The Collapse of PAL*'s narrative makes clear that the contemporary technosphere aesthetically discourages serious reflection on technological obsolescence and its wider implications, a discouragement that especially extends to the consumers responsible for the perpetuation of frictionlessness. The temporality of frictionlessness—a culmination of the violent storm of progress that blows Benjamin's Angel away from its care-inflected loyalties— and the aesthetic spectralization of its material effects induce a dismissal of matters of finitude and mortality. Even if the logic of optimization as the dominant ethos of perfectibility consists in nothing but a pervasive obsolescence that comes to haunt all things, frictionlessness prevents the formation of sustained chronolibidinal bonds and, instead, directs its users to a recurrent, destructive consumption. It deters a protracted temporality of care that would either mend and repair what is broken, that would question the purported undesirability of imperfect aesthetics and connectivity, or that would more carefully consider the fate of objects after they fall into disuse. Against this, both Puig de la Bellacasa's theorization of care and the investment in PAL that Menkman's performance invites disclose a Benjamin-like sensibility that challenges the logic of capitalist production and innovation by inferring new knowledges of the world and its destructions, illumined by an aesthetic attention for what is framed as outmoded and imperfect. The ongoing sense of responsibility that care implies delineates a drawn out and often frictional temporality of lingering and nurturing, of maintenance and repair, of contemplation, attention, negotiation, and devotion, and each of these dispositions is impeded by frictionlessness. While the temporalities of care are many, they are united in their tendency to disturb the accelerating times of production, innovation, and efficiency (Puig de la Bellacasa 2017: 210–11).

The drive for alternate temporal relations to technology that prove more attuned to finitude is echoed by anthropologist Shannon Mattern's broader suggestion that applying "'care' as a framework of analysis and imagination for the practitioners who design our material world, the policymakers who regulate it, and the citizens who participate in its democratic platforms...might [help us] succeed in building more equitable and responsible systems" (2018). Mattern offers a provocative thought here: what if we design technologies not primarily in view of the values of connectivity, user-friendliness, and optimization, but with an eye for care—care for the finite materiality of technology and care for those whose lives technology impacts? What if we design objects so that users sense rather than neglect the fragile and never infinitely exploitable resources on which they rely? What if we aesthetically integrate imperfection not as a dreaded prospect that urges disposal and replacement but as a resonant concept that can make one's engagement more enlivened? As *The Collapse of PAL* and *GlitchHiker* make clear, thinking with the concept of chronolibido and its aesthetic intensification provides one possible avenue for shaping more careful relations to technology and the worlds it sustains. Such artworks can affectively stimulate audiences to think more cautiously about how technological extinctions take shape.

It is, however, important to note that care also necessitates a thinking *carefully* about context. How a technology should and should not develop is always contingent on situational aspects and care is never *a priori* commendable; we cannot care about all things equally and so saving PAL from becoming a marginalized technological ghost might condemn other technologies to a silent afterlife. Indeed, Derrida urges us never to

hide from the fact that the principle of selectivity which will have to guide and hierarchize among the "spirits" will fatally exclude in its turn. It will even annhilate [*sic*], by watching (over) its ancestors rather than (over) certain others. [. . .] By forgetfulness (guilty or innocent, it little matters here), by foreclosure or murder, this watch itself will engender new ghosts. (1994: 109)

The lacuna of technological care that *The Collapse of PAL* identifies cannot be ameliorated by unconditionally welcoming all technological ghosts. There is no definitive answer to the conundrums the condition of autoimmunity and its always materially, historically, technologically specific specters throw our way. There are many possible responses when technological objects are at risk of breaking down or of being deemed imperfect. *The Collapse of PAL*, as a frictional performance that highlights the erasure of the outmoded, stages but one possible response. Steven Jackson's work on maintenance and repair (2014), discussed in the previous chapter, forms a related response, unveiling the hidden significance of restorative practices in the technological conduct of life. However, many contrasting but equally significant perspectives exist. Cultural geographer Caitlin DeSilvey's concept of curated decay, for example, foregoes repair and maintenance and finds a "radical potential [in] continued ruination" (2017: 90). She describes a host of examples where natural processes of decay and disintegration are allowed to take front stage. Within these curated sites of ruin, the principle of nonintervention has been elevated to a pedagogical gesture that emphasizes absence and vulnerability over continuity and control (DeSilvey 2017: 175). As another example, film scholar Matilde Nardelli criticizes naive dreams of a digital freedom from history and decay, and sees in the recent popularity of outmoded and imperfect objects in art exhibitions a different perspective: that of the old as needed to reinvigorate the new and therefore never truly obsolete (2009: 260).[7] These examples, diverse though they are, welcome rather than reject an aesthetic of imperfection, viewing it as something that might make its audience more aware of the material conditions of technological production. They enact divergent, care-infused responses to the situated effects the ineluctable primacy of imperfection produces and spell out that there is not one transcendent response that befits all situations. Ghosts will always arrive and return, but it is up to us to remain responsible for the technological shapes they inhabit.

Donna Haraway suggests that if "we cannot learn to live with ghosts," we "cannot think" (2016: 39). In a very real sense, these ghosts must include technologies. For one, because human consciousness is conditioned technologically, the result of a process of evolution that has taken shape through a lineage of now defunct but spectrally present technologies. Yet, as this chapter has emphasized, all the more because if we do not consider technologies as spectralizable forms that are central to the constitution of environments and societies, we cannot act responsibly. The type of silent execution *The Collapse of PAL* bemoans is never truly silent, but silent only

---

[7] Here, too, Benjamin proves oracular; much of his work was illuminated by the wager that objects started to reveal their hidden potential once they were "freed from the drudgery of being useful" ([1982] 2002: 19).

for some people—it is symptomatic of denying the materiality of ruin, of displacing consequences elsewhere, of comfortably hiding ghosts from view even as they catastrophically impact the world. Considering how technologies fade, what modes of living and dying they encourage or hamper, and for whom, is a necessary requisite for finding responsible relations in a world of collective, yet unevenly distributed finitude.

5

Ghostly Dreamworlds of Consumption

Cat System Corp., Vaporwave, and Tertiary Retention

The aesthetic of imperfection that defines the two objects I analyzed in Chapters 3 and 4 poses a critical counterweight to a cultural logic of incessant optimization and disposal. Behind this logic lies the issue of consumption and while I have thus far addressed this phenomenon only tangentially, in this last chapter I scrutinize a musical genre that directly interrogates the contemporary imbrication of technology and consumerism. The genre in question is *vaporwave*, an audio(visual) genre of electronic music that has garnered considerable online popularity over the years. Vaporwave frequently evokes connotations of both imperfection and spectrality in its listeners, and accordingly marks a prime point of interest for this study. In particular, I will focus on the work of Dutch vaporwave producer 猫 シ Corp. (henceforth Cat System Corp.). Through a close reading of three of his records, and particularly the record *Palm Mall Mars* (2018b), this chapter develops the argument that vaporwave, through an aesthetic of imperfection, formalizes what music writer Simon Reynolds terms the rhythm of *hyper-stasis*: a paradoxical amalgamation of technological acceleration and cultural standstill. Vaporwave enacts the effects this amalgamation has on human consciousness, a point I further explore by drawing on Bernard Stiegler's notion of *tertiary retention*. Ultimately, I argue that vaporwave's aesthetic of imperfection and its obsession with ghosts of the consumerist past dramatize a culture that cannot envision a future beyond the unfolding of ever more frictionless modes of consumerism. However, against this pessimistic vision, vaporwave also holds out the hope of new technological imaginaries that are less oriented toward endless consumption.

Vaporwave

The musical genre of vaporwave derives its name from the term "vaporware," which identifies (computer) products that were announced but never actually made it to the market.[1] Cultural theorist Laura Glitsos offers an apt introduction to the genre that I borrow here:

> [V]aporwave is a style of music collaged together from a wide variety of largely background musics such as muzak®, 1980s elevator music and new age ambience. The vaporwave song structure is usually short and repetitive, often slow (sitting around 60-90 bpm) with vocal samples positioned low in the mix saturated with heavy reverb and often slowed down to produce a "stretched out" effect or a "melting" quality. (2017: 100)

The collaged patchworks that Glitsos describes are seldom neatly sutured: vaporwave songs tend to be fraught with glitches, abrasive cuts, and unexpected deviations.[2] The content of these collages is sculpted from the perceptual experiences of "life under capitalism, both past and present" (Tanner 2016: 43); vaporwave's sonic palette draws from the raw material of capitalism's everyday detritus. For example, vaporwave frequently invokes technologies that capitalism's rapid turnover has rendered obsolete (or, as the previous chapter stressed, spectral)—VHS tapes, old operating systems, outmoded video game consoles. Moreover, it is the mass of mundane sounds encountered over the course of the everyday that harnesses vaporwave's aesthetic: 1980s and 1990s hits that play over supermarkets' speakers, the high-pitched sounds of card readers and ATM machines that punctuate commutes, and the tv/radio commercials whose low background hum accompanies people throughout the day.

Within this sonic plenitude, vaporwave shows a predilection for sounds designed to encourage consumption, and especially those sounds that evoke the technological quaintness of the 1980s and the (cyber)utopian dreams of the 1990s. The genre is digitally produced and mostly distributed online, often via online music platform Bandcamp, and the scene that has gathered around it cherishes anonymity, which makes it hard to tie this internet-born genre to a specific material context like a local scene or (sub)culture.

---

[1] These products thus retain a spectral form of being, enduring as lost futures that haunt the imagination without ever materializing—as Wendy Hui Kyong Chun claims, in support of her suggestion that "vaporiness is essential to new media": "Software . . . is ephemeral, information ghostly, and new media projects that have never, or barely, materialized are among the most valorized and cited" (2011: 21). Vaporware describes a situation where an emotional investment has been elicited that cannot be satisfied through consumption, as the product in question remains a vapory abstraction. Vaporwave, as I explain in this chapter, similarly evinces a stunted relation between consumption and gratification.

[2] Vaporwave's typifying techniques are not historically unprecedented and have numerous musical roots, such as the traditions of tape loop experimentation and field recordings, and the sample- and collage-based genre of plunderphonics.

This does not mean locality plays no role in vaporwave: the genre exhibits a persistent focus on the realm of the Western shopping mall, portrayed as a globalized phenomenon. The everyday nature of its sonic spectrum ensures that much of vaporwave's (often Western) audience has been subjected to its sampled signs for the better part of their lives, their childhood included, which largely explains the deeply nostalgic sentiments that surround it—nostalgia being one of the key conceptual lenses through which the genre has been analyzed (Tanner 2016; Glitsos 2017; Born and Haworth 2017). While vaporwave is a niche genre, its most popular albums have amassed millions of views on YouTube, and the avidity of its fan base suggests this music to reveal something fundamental about today's cultural landscape.

While I will mostly be concerned with vaporwave's musical content, it should be noted that vaporwave harbors an equally important visual component. This component corresponds with the genre's sample-based aural style: vaporwave visually juxtaposes (technological) objects from separate eras, toys with romanticized advertisement imagery, and relies heavily on computer graphics—sometimes the sleek aesthetic of today's digital technologies, but more regularly the pixelated, glitchy visuals that typified the personal computers of the 1990s. These elements are occasionally coupled with "Orientalist images of East Asia" (Born and Haworth 2017: 643) that serve as (retro)futuristic signifiers (Figure 5.1).[3] Vaporwave's collage-based appropriation of internet aesthetics has ensured that it exists not only as a musical genre but also as something of a "visual meme" that has been widely popularized and repurposed (Killeen 2018: 626). While this visual dimension is worthy of analysis in its own right, often even existing separately from the music, in this chapter I focus mostly on vaporwave's musical qualities, as it is here that I argue the genre's aesthetic of imperfection is most pronounced.

Vaporwave, Imperfection, Spectrality

If "the beauty of vaporwave lies in its imperfection" (Jury's Out 2017), then how does this imperfection sonically and thematically manifest itself? As the preceding passages already imply, vaporwave's digitally instantiated aesthetic of imperfection is characterized primarily by three elements:

---

[3]   This point also explains why vaporwave producers often use Kanji typography for their artist name and for their album and song titles. Cat System Corp. is a prime example of this practice; his artist name is generally transcribed as 猫 シ Corp.

Figure 5.1 The front cover of Cat System Corp.'s album 太陽。慰安。楽園 (2016) is an example of a typical vaporwave cover, combining idealized imagery of a beach resort with outdated, crudely assembled computer graphics and Kanji typography. These various elements are anachronously accompanied by a Roman pillar. Image source: f4.bcbits.com/img/a3588266587_16.jpg. Accessed October 31, 2022.

a) a heavy use of glitches, cuts, loops, and other techniques that prevent smooth linear progress;

b) an abundance of acoustic effects that synergistically refute clarity, such as echo, static, reverb, and distortion;

c) a musical preoccupation, despite the genre's digital nativity, with outmoded, often faltering technologies and obsolete media aesthetics.

The way vaporwave's imperfections take shape is akin to the aesthetic of imperfection that marks the objects I analyzed in previous chapters. Like *GlitchHiker*, vaporwave draws heavily on glitches to refute ideas of technological permanence. Like *The Collapse of PAL*, the genre displays an anachronistic aesthetic that paints a technologically disjointed temporality.

The notion of a time out of joint forms one of the main constituents of the spectral and, within vaporwave, the figure of the ghost is indeed never far off, as is also evinced by the title of cultural critic Grafton Tanner's 2016 book *Babbling Corpse: Vaporwave and the Commodification of Ghosts.*

Vaporwave, first of all, heavily employs the technique of collage. As music writer Simon Reynolds explains, "the sample collage creates a musical event that never happened; a mixture of time-travel and seance. Sampling involves using recording to make new recordings; *it's the musical art of ghost co-ordination and ghost arrangement*" (2011: 313–14, emphasis added). Vaporwave, moreover, often draws on medial forms that the previous chapter theorized as spectralized: technologies that have been overtaken by more frictionless alternatives while remaining culturally and materially present. The disjointed time that vaporwave's conjurations stage is made all the more forceful by a reliance on reverb and delay as acoustic effects that further imply the workings of the temporal. Furthermore, in line with the ghostliness of glitch as discussed in Chapters 3 and 4, vaporwave's glitch-based aesthetic of technological failure and error invokes the quality of glitch to point to technological death and to reveal unruly technological powers—see, for example, vaporwave producer Infinity Frequencies' trilogy of albums *Computer Death* (2013), *Computer Decay* (2014), and *Computer Afterlife* (2014). Specters also occur in meta-reflections about the genre's vitality: although still highly popular, already from its inception the genre has often been declared critically dead, lingering as a ghost long stripped of its subversive energies—see, for example, producer Sandtimer's 2015 album *Vaporwave is Dead*.

These are but some of the attributes that explain vaporwave's spectral connotations (cf. Tanner 2016: 5–13, 33–9). However, the most prominent spectral quality of vaporwave is its exhibition of a hauntological aesthetic. Hauntology, as I explained in Chapter 2, describes an aesthetic, imperfection-oriented approach to technology that explores how perception, memory, and anticipation are mediated by technology, allowing the lost futures of the past to clash up against the present. I will return to this ghostly quality of vaporwave later in this chapter and will also discuss how it relates to the notion of technological spectrality, but first I zoom in on the (music-)cultural logic behind vaporwave's aesthetic of imperfection.

As with my previous case studies, the technological aesthetic of imperfection under scrutiny here takes shape against the philosophy of frictionlessness that pervades contemporary technological culture (see Chapter 2). On a formal level, within the context of music production, the effects of this philosophy can be discerned in the normative "paradigm of digital perfectionism" (Strachan 2017: 136) that characterizes the many software plug-ins and recording technologies that allow today's producers to smooth over flaws and remove traces of mediation from the recording process. The result of these manifold applications is that we now witness a "digitally constructed notion of perfection in contemporary recordings" (Strachan 2017: 152), or, phrased differently, a standardization of digital interfaces encouraging production that tends toward the "mechanically perfect" (Kelly 2009: 271). Vaporwave, owing to the slight fog in which it shrouds its sounds and to the jarring transitions it refuses to conceal, is one digitally born genre that challenges the airbrushed mediations of the contemporary mainstream. It is no incident that vaporwave draws heavily from *muzak*, that most frictionless form of music, composed precisely to form the imperceptible yet expedient undercurrent to the navigation of commercial and corporate space—it is "the lubricant that glides us along our journey of daily material existence" (Tanner 2016: 40). Muzak is a material condensation of the value of user-friendliness I discussed in Chapter 2: stripped entirely of the frictional peculiarities of personality and content, it makes no demands and imposes nothing, merely seeping around the thresholds of perception to ameliorate tension and smoothen commercial transaction. It is music reduced to a technical function. As will become apparent, vaporwave's musical significance lies in how it highlights the friction in the frictionless by aesthetically wresting a feeling of tension from familiar tokens of techno-consumerist culture.

Before corroborating these claims and introducing my prime object of analysis—the work of producer Cat System Corp.—it is important to first consider two albums that have been seminal to vaporwave's development. While these two albums are surely not the only two objects to inform vaporwave's aesthetic, they merit discussion here because they display two different temporal models that, in conjunction, exemplify vaporwave's unique rhythm of what Simon Reynolds has termed hyper-stasis. Situating Cat System Corp.'s work against the background of these canonical albums will substantially deepen our understanding of vaporwave.

Eccojams Vol. 1, *Far Side Virtual*, and the Rhythm of Hyper-Stasis

In December 2019, the eclectic music webzine *Tiny Mix Tapes* published a list of what they considered the 100 best albums of the decade. Chuck Person's *Eccojams Vol. 1* (2010) was awarded first place, a testament to the record's cultural resonance. Chuck Person is one of many aliases of producer Daniel Lopatin, who is best-known for his ongoing work under the name of Oneohtrix Point Never. The record's eponymous term "eccojams" (consonant with "echo jams") refers to a set of techniques that serve as progenitors of the vaporwave sound. Eccojams are isolated snippets of pop-cultural material that have been chopped up, slowed down, looped, glitched, and veiled in reverb and static. These samples have, in other words, been actively subjected to a technological aesthetic of imperfection that agitates against the aesthetic of perfection and frictionlessness described earlier. Lopatin began releasing such eccojams on YouTube under the username sunsetcorp[4] and later gathered a selection of them on *Eccojams Vol. 1*, featuring, for example, "eccojammed" renditions of Toto's "Africa" (1982) and Janet Jackson's "Lonely" (1989). The result is a ghostly vignette of the pop-cultural past that shapes an atmosphere of anachrony similar to the work of some of the hauntological musicians mentioned in Chapter 2. Perhaps the best example of the peculiar quality of Lopatin's eccojams is the song "A3," which features a reworking of Ian van Dahl's "Castles in the Sky" (2002), a classic from the trance genre. The pristine original has been slowed down and saturated with echoes while a sense of destitution bleeds into the vocals, and the song has been stripped of its communal spirit. Whatever shared rushes of dance-floor euphoria it originally inspired have been dispelled and the tacit promise of endless nights so prevalent in club music has been broken. The celestial castles to which the song refers now appear more like ruins, their halls reverberating only with voices from the past and the hazy memory of strobe lights. Certainly, the song retains some of its original blissfulness—but this bliss is refracted inward instead of toward a horizon of collective catharsis.

This subversion of songs from the past and particularly of the emotions associated with them is what characterizes the general atmosphere of *Eccojams Vol. 1*. The record comprises a collection of warped and decelerated pop songs from different eras (Jojo's 2006 hit song "Too Little Too Late" sits alongside Marvin Gaye's 1982 hit "My Love Is Waiting") whose uplifting momentum and radio-ready sheen have been exorcized—the hopeful vision of the world the original songs offered appears presently in splinters. This strategy is redoubled by Lopatin's tendency to isolate lyrical particles that, upon repetition, radiate solitude and inertia ("there's nobody here" on "B4;" "just one more year and then you'll be happy" on "B3"). The most effective eccojams thus wrest friction from songs that were composed to feel decidedly carefree and optimistic.

This frictional effect imparts the theme of lost futures so germane to hauntology. In Lopatin's work this plays out not only on the level of the song but also as a commentary on the cultural logic behind it. As Simon Reynolds argues, "the surplus value Lopatin brings is the conceptual framework to his projects, which relates to cultural memory and the buried utopianism within capitalist commodities, especially those related to consumer technology

---

[4] See www.youtube.com/user/sunsetcorp (accessed February 4, 2021).

in the computing and audio/video entertainment area" (2011: 81). The act of excavating such spectralized, utopian futures defines the work of many vaporwave producers, invested as they are in revisiting ordinary, often outmoded signs of commerciality such as commodified pop songs. By framing the futures these signs once promised in a broken light, they disclose the different temporalities contained within the mundane.

Crucially, Lopatin's eccojams and the genre of vaporwave as a whole cannot be understood as only an expression of or exercise in nostalgia. In conversation with Reynolds, Lopatin emphasizes this point: "'I'm super into formats, into junk, into outmoded technology,' says Lopatin. 'I'm super into the idea that the rapid-fire pace of capitalism is destroying our relationship to objects. All this drives me back, but what drives me is a desire to connect, not to relive things. It's not nostalgia'" (quoted in Reynolds 2011: 83). Lopatin, in a rationale that resounds with *GlitchHiker*'s facilitation of a profound objectual attachment (Chapter 3) and Rosa Menkman's conjuration of Walter Benjamin's Angel of History (Chapter 4), denounces the destructivity of progress and sees his engagement with the past not as a form of revisionism, but as a possible invention of new temporal imaginaries. Specifically, his reconfiguring of the musical loop—as a "cultural form" that condenses "capitalism's reliance on repetition, familiarity, and virality" (Jackson 2019)—shows that pop culture's commodified signs can always be repurposed, sensed, and experienced in new ways. As music journalist Marvin Lin writes, in a strongly Derridean spirit, *Eccojams Vol. 1* demonstrates that "nothing is truly fixed and everything is ripe for transformation" (2020: 170). The record thus armed vaporwave with a template that tied imperfection-oriented techniques (looped and slowed-down samples, warped vocals, glitches, and a heavy use of echo and static) to a conceptual register that pertains to consumer capitalism.

A second album that has been crucial to vaporwave's development is James Ferraro's *Far Side Virtual* (2011), often abbreviated to *FSV*—an album that Ferraro himself at the time described as "the darkest record [he had] made in [his] life" (Bowe 2013). James Ferraro is an American producer who began his career in the genres of ambient and drone, but who has since branched out into numerous other genres. *FSV* can best be qualified as an artificial high-resolution picture of everyday life under a thoroughly digitalized capitalism. As Ferraro states, the record aims to grant a momentary "whole view" of society and to capture the general rhythm and affects of contemporary existence (Gibb 2011). "*FSV*," in his words, "is plastic, pseudo-Utopian and lifeless sounding, but co-exists with the darker aspects of reality" (Gibb 2011). It is telling that Ferraro initially planned to release the music as a series of ringtones; modern life as Ferraro paints it is wholly pervaded by information technology and its subjects always exist proximate to digital devices ready to link them into globalized networks of consumption. On the whole, *FSV* captures the feeling of a world regimented by the core values of frictionlessness: a culture quickened by the demands of user-friendliness, connectivity, and optimization.

The album's musical content can be described as an upbeat kaleidoscope of corporate jingles that has been designed to lack distinct personality. It enacts an accelerated version of shopping-mall muzak, and the album's motley of saxophones, pianos, and sitars sounds like presets from a cheap Casio keyboard. The intersection of consumerism and digital technology is accentuated through spoken samples ("Sir, would you like to receive the New Yorker directly on your iTablet?" on "Global Lunch"; "Take a look at our virtual sushi menu" on "Palm Trees, Wi-Fi and Dream Sushi") and through song titles ("Fro Yo and Cellular Bits"; "Starbucks, Dr. Seussism, and While Your Mac Is Sleeping"). Ferraro states that he sought to draw attention to the multitude of sounds that surreptitiously serve "as a social lube for capitalistic transactions," revealing a hegemonic culture of "corporate consumer fetishism mixed with American nostalgia" (Grunenberg 2021). Conjuring the spectral, Ferraro explores the tension of a still prevalent consumerism now confronted by "the decline of American prosperity, a ghost of a once-superpower that is dying" (Grunenberg 2021), less and less able to offer people an attainable image of the good life.[5]

FSV enacts the sense of traversing a shopping district, where sped-up muzak commingles with the sounds of ATM machines, mobile phones, and other digital interfaces, while soda machines ceaselessly pour high-fructose energy drinks. Its feeling of constant action and enterprise makes *FSV*, above all, a *zany* record in the sense that cultural scholar Sianne Ngai has theorized this term. *FSV* exhibits precisely the contemporary fetishization of activity and movement that zaniness captures—it depicts a "blur or stream of undifferentiated activity" that ultimately feels more exhausting than empowering (Ngai 2012: 202). Ngai theorizes zaniness as an aesthetic of movement that is indexical to labor and production, but *FSV* shows

---

[5] This focus on a stunted relation between consumption and satisfaction links *FSV* and vaporwave more generally to theorizations of the tension between the promises of consumption and the reality of widespread economic turmoil. Think, for example, of Henri Lefebvre's conceptualization of the alienation that ensues when a focus on fulfillment becomes "the principle of one's decline and loss" (2014: 508) and of Lauren Berlant's more recent notion of cruel optimism as a relation that "exists when something you desire is actually an obstacle to your flourishing" (2011: 1). Such relations, Berlant argues, are particularly pervasive today, now that the consumerist good life that capitalism continues to promise is proving less and less tenable.

that zaniness can be a consumption-driven aesthetic as well. As indicated previously, vaporwave distills friction from the frictionless, and this is precisely what *FSV* accomplishes—music writer Miles Bowe trenchantly quips that the record is "pleasant to the point of nihilism" (2013). *FSV* indeed mirrors a society designed to be absolved from friction, a digitalized world still tightly tethered to the quaint and lively optimism of unbridled consumption. Yet, the record's zany antics and its backdrop of economic recession lend its forced smile a sociopathic visage, recalling one to Ferraro's description of the record as the darkest in his career. Ultimately, the record's strenuous exertions accommodate as much of a tragic undertone as the zany and desperate movements of Ngai's odd-jobbers and precarious laborers do (Ngai 2012: 188).

The respective rhythms of *Eccojams Vol. 1* and *FSV* could hardly be more different—the anemic drawl of Lopatin's loops vis-à-vis Ferraro's zany rendition of an accelerating modern life. Yet, it is from the unlikely coupling of these distinct temporalities that vaporwave's ostensible ethos emerges.[6] More concretely, the temporal logic most specific to vaporwave is the rhythm of what music writer Simon Reynolds terms *hyper-stasis*. Reynolds characterizes this rhythm of technological velocity and cultural inertia as follows: "In the digital present, everyday life consists of hyper-acceleration and near-instantaneity (downloading, web pages constantly being refreshed, the impatient skimming of text on screens), but on the macro-cultural level things feel static and stalled. We have this paradoxical combination of speed and standstill" (2011: 427). Hyper-stasis thus denotes a situation where technology accelerates and becomes more frictionless only to stultify cultural and humanistic progression; it bespeaks a stunted culture that clings to its past rather than inventing new futures. It is an acute magnification of the logic of optimization: even though technology constantly updates and accelerates, the underlying cultural rationale of constant, repetitive consumption remains unaffected, paradoxically leading to a situation where

---

[6] Lopatin's and Ferraro's opposing rhythms coalesce in what generally counts as the canonical vaporwave record: Macintosh Plus's フローラルの専門店, or *Floral Shoppe* (Macintosh Plus is an alias of prolific vaporwave producer Ramona Andra Xavier). While a full analysis of this album exceeds the present chapter's ambit, it warrants mentioning here and is particularly noteworthy for how it mixes the slow temporality of *Eccojams Vol. 1* with the technological inclinations of *FSV*, creating the sensation of a glitchy simulator designed to scrape one's remembrance of halcyon days. The song リサフランク420 / 現代のコンピュー is perhaps the most well-known vaporwave track, reimagining Diana Ross's 1984 hit song *It's Your Move* as a confounding palette of impromptu cuts and spectral voices. Also noteworthy is the economic terminology that characterizes the song ("I'm giving up on trying to sell you things that you ain't buying"), which already hints at vaporwave's sustained fixation on consumerism.

ceaseless novelty engenders reiteration, stasis, and nostalgia. Hyper-stasis indicates, in short, a time out of joint with itself.

I draw on Reynolds's concept primarily because he minted it in the context of music, but it is but one among many theorizations that reveal the present to be punctuated by a temporal disjointedness. Three philosophers who have canonically charted the cultural impasses produced by technological acceleration are Henri Lefebvre (2014), Fredric Jameson (1991), and Paul Virilio (2006). More recently, there is Hartmut Rosa's critical concept of "dynamic stabilization" (2019: 21); Bernard Stiegler's characterization of technological innovation as a force "that changes everything in order that nothing changes" (2019: 191); and Thijs Lijster's suggestion that "the central paradox of our time" concerns how the "experience of accelerating life goes hand in hand with the experience that nothing *really* changes" (2018: 221, emphasis in original). Eliding their respective differences, these authors all contend that technological optimization and acceleration are certainly not the same as, or are even inimical to, cultural and social progress. Similar arguments have also been set forth by Mark Fisher and Franco "Bifo" Berardi, to both of whom I return later in this chapter. Next to Reynolds's definition, the term that best captures this widely theorized tension of stasis and action is Wendy Hui Kyong Chun's phrase "updating to remain the same," which delineates a technological disposition of constant optimization and updating that produces no real forward momentum, trapping users in a "never-advancing present" (2016: 76). These paradoxes of stasis and velocity are reflected in the juxtaposition of Lopatin and Ferraro's work. *Eccojams Vol. 1*'s temporal model consists in inertia, symbolizing cultural malaise through a deceleration of commodified pop songs. *FSV*, on the other hand, portrays cultural calamity through an accelerating logic of technological connectivity and consumption.

In short, the rhythm of hyper-stasis is a rhythm that reveals the friction in the ostensibly frictionless, the imperfect in the purportedly perfect; it complicates optimization- and velocity-bound trajectories of technological perfection by integrating a rhythmic underside—accentuated by deceleration, loops, and glitches—of psychopathological and cultural inertia. Keeping in mind this temporal template and vaporwave's general preoccupation with technology and consumption, I now turn to this chapter's central object of analysis: the work of Cat System Corp.

Cat System Corp.

Cat System Corp. is an influential vaporwave producer who originates from the Netherlands and whose real name is Jornt Elzinga. Like many of his vaporwave peers, Cat System Corp. is highly prolific: as of April 2021, he has released at least thirty-seven albums and twelve EPs, with the first album dating from 2013. He also runs the record label Hiraeth—"hiraeth," significantly, is a Welsh word that describes a longing for a place that never was. This name connotes the peculiar sense of a time unhinged that haunts vaporwave, where meditative reveries are always tinged by the realization that the content of one's memory may be a retrospective fabrication. Cat System Corp. draws on the vaporwave aesthetic to address a wide number of themes, ranging from contemporary digital technology (in, for example, his 2018 release *A Class in . . . CRYPTO CURRENCY*) to a nostalgia for 1980s television (the topic of 2015's *Class of '84*). His work is primarily of interest for how, contrary to the output of many vaporwave producers (whose dispositions are sometimes difficult to distinguish from nostalgia), it employs a productive sense of hauntology that complicates the cultural attachment to tokens of consumerism. I first discuss how a preoccupation with consumerism, media aesthetics, and cultural flux emerges from some of Cat System Corp.'s most renowned records, after which I move on to examine *Palm Mall Mars* (2018b), a record that epitomizes Cat System Corp.'s aesthetic and that forms the gist of the analysis I develop in the second half of this chapter.

The motif of a troubled or ambiguous relation to consumerism is most dramatically communicated by *NEWS AT 11*, one of Cat System Corp.'s most acclaimed works. This record, released on September 11, 2016, places its audience in a simulated reality composed of actual weather channel samples and radio outtakes that were broadcasted right before the first plane crashed into the World Trade Center on September 11, 2001. The record refuses, however, to face this event head-on and consistently draws away right before the moment of impact: the album distinguishes itself through a constant sense of deferral. This is obvious from the outset of the record, when *Good Morning America* co-hosts Charles Gibson and Diane Sawyer cheerily announce that it is "Tuesday September —," the sound fading away before they have a chance to utter the number "eleven." *NEWS AT 11* is littered with these kinds of eerie-in-hindsight messages—on "Financial News," a reporter looks ahead to a "solid open to the trading day" and ominously mentions Boeing stocks; on "Heli Tours," a journalist implores us to "look at Washington" on this "perfect September day" ("I'm going outside today!"); and on the track "8:46AM" (the minute the first plane hit), a reporter tells us that "it's kind of quiet around the country—we like quiet." The radio advertisements that *NEWS AT 11* samples similarly transmit an effervescent mood that becomes frictional in light of what was to happen moments after their transmission. The track "Financial News," for example, features the McDonalds commercial ("we love to see you smile") that was notoriously the last ad aired before NBC commenced its live coverage of the events. These reconfigured media traces are supplemented with a selection of smooth and uplifting jazz tunes that have been subjected to the familiar vaporwave treatment—decelerated, doused in reverb, cut up, and suspended on thin layers of tape hiss. The most powerful example of the tension this generates comes at the end of the track "Tuesday Television," when we hear *Today Show* host Matt Lauer cut his interview with writer Richard Hack short to switch to footage of 9/11's unfolding events. Instead of making listeners privy to what happens next, *NEWS AT 11* shifts away and segues into the track "Evening Traffic," which consists of a sample of Dan Siegel's "Where Are You Now?" (1984) that has been manipulated in the vein of *Eccojams Vol. 1*. The slowed-down pace and heavy echoes give the song's melody a deeply melancholic and plaintive quality, which is made all the more potent by the record's concept and the track's placement right after the interview.

The events of September 11 thus constantly haunt the record while never gaining full form. *NEWS AT 11* keeps invoking the time right before the crash, painting a picture of a stunted culture that is forced to revisit time and again the reality of rupture, each time looking away and regressing into denial. Through coupling the events of 9/11 with recurring mantras of cheerful consumption and economic prosperity, *NEWS AT 11* dramatizes a rhythm of hyper-stasis. On the one hand, the album hounds the listener with cheerful advertisements and weather forecasts, and with sampled jazz tracks of a lighthearted and easy-listening variety, all suggesting forward momentum. On the other hand, it courts disorientation, not only by constantly cutting between samples and by sonically mobilizing a reverb- and static-based aesthetic of imperfection, but also because the American culture of cheer that it portrays unravels in view of the horror that is persistently deferred yet spectrally present. The most obvious reading of *NEWS AT 11* would be one through the lens of historic trauma[7] and the cultural inability to work through it, but I present the record here to emphasize another aspect. I posit that by invoking the mediatic articulations that directly preceded a disastrous event, *NEWS AT 11* paints an extreme instantiation of the notion that consumption, for all its promises and expectant projections, has far from inaugurated a carefree utopia. The record's significance, in other words, certainly lies in its focus on 9/11 as a singularly traumatic and terroristic event, but also in its problematization of the joys of consumption. By focusing so intently on the optimistic

---

[7] On vaporwave and trauma, see, for example, Glitsos (2017: 107–10) and Tanner (2016: xi). For further reading on the relation between 9/11 and vaporwave, see Tanner (2016: 51–4).

media transmissions that prefaced a far-reaching moment of catastrophe, Cat System Corp. makes their upbeat atmosphere appear f(r)ictional and estranging.

This ambivalent relation to consumption informs most of Cat System Corp.'s work, as is made clear by the so-called genre of mallsoft of which many of his records are part. Mallsoft is a sub-genre of vaporwave that evokes the sense of wandering aimlessly and endlessly through a shopping mall, pleasant but also vaguely unsettling; while following *FSV*'s preoccupation with a zany consumerism, the genre's drifting aesthetic is often more in line with *Eccojams Vol. 1.* The mall has an unequivocally central status within vaporwave—it is no incident that the online Reddit forum dedicated to vaporwave carries the adage of "music optimized for abandoned shopping malls" (note the ironic repurposing of the value of optimization). In vaporwave, the shopping mall shines as an apex of consumerism and cultural uniformity: malls are envisioned as brightly lit idylls whose corridors, plazas, and water fountains are designed to smoothly assimilate the visitor into the perpetually expanding networks of global capitalism.[8] The appearance of the Western shopping mall disregards local idiosyncrasies, boasting a monotone aesthetic that is immediately recognizable.[9] It is also the place where culture can be flattened: the mall is a site where punk records can sit comfortably next to softrock albums, and a cornucopia of cultural specificities can be crammed into the unifying confines of the discount bin. One of the great specters that stalks the pages of Derrida's *Specters of Marx* (1994)—the specter of exchange value—reigns here, forcing all of the mall's attributes to service the flow of capital. As a genre, mallsoft intensifies vaporwave's already distinct preoccupation with consumerism and the mall: it is steeped in "directionless and echo-heavy samples" (Chandler 2016) that create the experience of drifting absent-mindedly through one of these commercial behemoths. Alongside Cat System Corp., some of mallsoft's main producers are Disconscious, Hantasi, 鬱 [Depressed], and 식료품 [groceries], and each of these artists weaves alternately (faux)-utopian and dystopian dreamscapes from a curation of mall-related samples.

---

[8] Adam Harper, one of vaporwave's most influential exegetes, has written about vaporwave's relation to the mall as (virtual) plaza, but his analysis fails to address why, if the mall is such a bristling place of consumerist "activity" and "spectacle," vaporwave's malls feel so unspectacular and eerily devoid of human agency (2012).

[9] For more on such ambiguous and monocultural spaces designed for consumption or transit, see Marc Augé's famous 1992 work on *non-places* ([1992] 2007), a term he uses to refer to indeterminate and anonymizing places like shopping malls and hotel lobbies. For further reading on the history of the American image of the shopping mall, see Howard (2015); Scharoun (2012).

Cat System Corp.'s 2014 record *Palm Mall* counts as a classic of the mallsoft genre and is probably his most well-known work. Through some of the usual components of vaporwave's aesthetic of imperfection, most notably the slowing down of samples and a liberal use of reverb and echo, this record encapsulates mallsoft's central atmosphere: a constant oscillation between the oneiric and the eerie—between capitalism's promises and its ghostly outcomes. Mark Fisher has theorized eeriness as an affect that congeals around vacant structures or scenes that have otherwise been emptied of human presence (2016: 11). Eeriness evokes the impression that there are unbidden forms of agency at work that escape the grasp of the beholder (2016: 11). That the eerie is often associated with serenity (2016: 13) makes intelligible how mallsoft's dreamy tones can so easily slide into more troublesome territories; something often does not feel quite right about the tranquil malls that mallsoft depicts.

Palm Mall's 22-minute[10] titular track epitomizes mallsoft's peculiar eeriness. This track is, in fact, more of an ambient piece than a proper song, steering the listener past a wealth of washed-out fragments of muzak and multilingual advertisements that underline the ubiquity of capital. The track's sounds are sampled from both actual malls and virtual ones—the piece includes a sample of the North Point Mall that features in the 2002 video game *Grand Theft Auto: Vice City*, blurring the lines between the virtual and the "real" when it comes to capital's reach. While not entirely stripped of human presence (the faint sounds of human merriment occasionally ingress into the scene), the dreamlike air suggests that true agency is located elsewhere; with track titles like "I c o n s u m e , t h e r e f o r e I a m" [double spacing in original] and "Veni Vidi Emi" ["I came, I saw, I bought"] the album's focus on docile consumerism and its implications of existential meaning through commercial purchase is obvious. Even though humans are present, a sense of eeriness emerges, as the album's hazy aesthetic presents these humans more as lingering vagrants than as vigorous beings in full command of their route through this outpost of capital. Significantly, Fisher designates capital as an unparalleled entity of eeriness—a powerful and ghostly abstraction that endlessly yields material effects, ensuring that "'we' 'ourselves' are caught up in the rhythms, pulsions and patterning of non-human forces" (2016: 11). *Palm Mall*'s hyper-static take on *FSV*'s zany consumcrism dramatizes this suggestion by envisioning subjects that seem spellbound by the promises of consumption, drifting through malls that appear more as ghostly mirages

---

[10] The title track's length is highly unusual for vaporwave, as the genre is generally characterized by short tunes that seldom exceed the four-minute mark.

than as actual, physical places capable of delivering their visitors to a better, more prosperous tomorrow.[11]

In 2018, Cat System Corp. complemented *Palm Mall* with a spiritual successor, *Palm Mall Mars*, and it is this record that I argue to encapsulate the full stakes of vaporwave's hauntological aesthetic of imperfection. *Palm Mall Mars* protracts *Palm Mall*'s mallsoft project with one important twist: the record's eponymous mall is located on a future planet Mars. The album comes with the following description:

> Mankind cheered: not even 100 years after the first human set foot on the moon we made our first Mars colony.
>
> Another 100 years later we cheered again: the first craters on Mars have been successfully terraformed, cities are built and mankind is ready for its first migration to another planet. To celebrate Earths Senate CEO opened in 2149 the first Martian shopping mall: PALM MALL MARS!
>
> Today, the year 2199, we celebrate the 50th birthday of Palm Mall Mars. We welcome you with special discounts on luxurious items, grand offers on newly built Ring Worlds and will let you try out the new ARPE! Come visit us and get a 50% discount on your first Poulsen Treatment!
> > See you at Palm Mall Mars! (Geometric Lullaby 2018)

What immediately stands out when listening to *Palm Mall Mars* (Figure 5.2) is that it is not all that different from the general mallsoft aesthetic and from *Palm Mall* in particular, even incorporating some of the same sounds. Again, we find an assortment of hazily rendered new age/jazz tunes with twisted vocals that reverberate through the mall's spacious halls. Again, there is the eerie sense that our agency has been thrown into question. Signs of consumption are everywhere: the song "Second Floor" incorporates product scanner-sounds akin to those of the first *Palm Mall* album; "Poulsen Treatment Studio" features a voice that orders us to "insert cash or select payment type"; and on multiple tracks, PA advertisements ring in the distance. At the end of the album, we are notified of the imminent departure of our flight back to Washington, which presumably puts an

---

[11] *Palm Mall*'s second half is more conventionally structured, featuring individual tracks of reimagined pop tunes that are interspersed with advertisements and the sounds of product scanners, further suggesting capital's surreptitious presence. The record's shift from ambience to a more upbeat sound mirrors what media scholar Jonathan Sterne identifies as a central tensional quality of the mall's acoustic space: "[Q]uiet nondescript music in the hallways contrasting with louder, more easily recognizable and more boisterous music in the stores" (1997: 29).

Figure 5.2 *Palm Mall Mars*'s front cover juxtaposes the familiar non-place of the escalator with an image of virtualizing technology. Image source: https://f4.bcbits.com/img/a3505265934_10.jpg (cover design by Cat System Corp. and Christopher Hansen). Accessed October 31, 2022.

end to this interplanetary voyage, but there are few other signs that might alert one to the radical leap in time we have supposedly taken. Consider, for example, "Poulsen Treatment Studio," whose joyous sax melody is so shrouded in tape hiss that it feels more as if one is listening to a timeworn tape machine than to a recording from a future that would surely feature more high-fidelity means of transmission. Similarly, according to the liner notes, the song "홀로 그래픽 컴패니언 ARPE" references a state-of-the-art technology, but the song's glacial pace and pitch-shifted vocals sound decidedly a-futuristic and lackluster.

What is, in other words, remarkable about *Palm Mall Mars* is that, while it purportedly takes us to the year 2199, we still find ourselves traversing an imaginary mall that is effectively an amalgam of 1980s/1990s imagery. Time seems thoroughly out of joint: how can the far future feel so familiar, archaic even? One reviewer criticizes *Palm Mall Mars* for staying too close to mallsoft's established tropes (DreamWorks), but I propose this is exactly the point: *Palm Mall Mars* charts a future in which a technological culture that has mastered the frontier of space travel only sees humankind importing the model of the shopping mall to other planets. Technological innovations notwithstanding (*Palm Mall Mars*'s track titles mention indoor climate control and a virtual world generator), this galactic environment is still entirely routed and undergirded by a logic of consumerism. *Palm Mall Mars* is an extreme aesthetic enactment of Fisher's vision of "[t]he arid shopping mall at the end of history [that has] opened up as the only possible future" (2018b: 299). As far as vaporwave records go, few albums capture so adequately the rhythm of hyper-stasis: technological acceleration as tied to a cultural sensation of drift, where the mall's dwellers are perpetually promised novelty but appear for centuries to have been deprived of the shock of the new.

Which brings me to a last important consideration about the role of the mall in vaporwave: the fact that, contrary to the suggestion of a shopping mall that awaits us at history's end, the physical realm of the shopping mall is, in many places, in decline. For some years, we have been witness to what industry experts describe as a retail apocalypse, or the widespread disappearance of retail stores, primarily across the West (Howard 2015; Pilkington 2019; Helm, Kim, and van Riper 2020).[12] The reasons for this disappearance are many—the 2008 crisis and massive company debts being two among them—but one of the most obvious causes is the escalation of the philosophy of frictionlessness and the realm of e-commerce. Frictionlessness is geared toward the disintegration of obstacles that might prevent consumption, aiming to restrict "*différance* to the shortest possible circuits in order to gain the most rapid possible return on investment" (Stiegler 2017: 51). From this perspective, the physical mall is a potential obstacle that is to be dissolved in the lightning-fast networks of the internet. Symptomatic of this rationale is, for example, the ever-increasing capacity of Amazon's Echo products, whose virtual assistant functionalities allow users to order items through voice interaction, cutting out the role of the shopping mall entirely. The effects of such technologies on the territory of the shopping mall are clear. Market expert Greg Guenthner deduces that the decline of the shopping mall has been thoroughly accelerated by Amazon's quest to condemn to obsolescence "every brick and mortar store in their path" (2016). The mall's downturn thereby adds another spectral layer of significance to vaporwave's preoccupation with soporific, eerie,

---

[12] This phenomenon underlines, despite its claims to universality, the Western nature of the genre. In China, for example, outlet malls are actually on the rise, despite the ubiquity of e-commerce (Arcibal 2019). I thank Alex Williams for pointing this out to me.

and abandoned shopping malls,[13] because vaporwave artists document an experience whose prime is arguably behind them.[14]

Vaporwave and Tertiary Retention

What to make of this concerted focus on the mall in a time of frictionless consumerism? Do, moreover, the discussed records, and the genre of vaporwave more generally, genuinely yearn for the carefree consumerism the mall promises or do they seek to unsettle such fantasies from within? I do not aim to settle the question of authorial intent here. Rather, what I want to focus on is what vaporwave's aesthetic framing of the mall as a site of both desire and eeriness says about the technical conditioning of memory, perception, and anticipation within consumerist societies. That vaporwave invites both associations of pacified comfort and cultural critique is explained by its ability to speak to highly individual recollections of growing up in a world saturated with commercialized technological signs while at the same time registering the immanence of these signs to the collective; vaporwave "brings to mind a shared cultural memory" and "reflects upon processes of remembrance in the context of post-digital mediation" (Strachan 2017: 146). More specifically, I propose that vaporwave is, at heart, and certainly not always wittingly, a genre about a certain prevailing composition of *tertiary retention*.

Tertiary retention refers to the exteriorized, technical element of consciousness that Bernard Stiegler theorizes as constitutive of human perception and cognition. While I already addressed this concept in Chapter 1, it is worth to briefly reintroduce it. In developing the notion of tertiary

---

[13] Tellingly, Dan Bell's ongoing *Dead Mall* series on YouTube serves as inspiration to many vaporwave producers. In this series, Bell gives his viewers tours of American malls that have been largely abandoned in the wake of capitalism's migration to the internet. There is an eerie and haunting quality to these videos, the malls appearing effectively as ghosts of their former selves, harkening back to a time when, supposedly, their empty halls and stores were still pregnant with cheer. See https://www.youtube.com/playlist?list=PLN z4Un92pGNxQ9vNgmnCx7dwchPJGJ3IQ.

[14] Here, vaporwave veers closest to the previously discussed work of Walter Benjamin. Benjamin charted the decline of the Parisian arcades—sites that once offered their visitors a dreamworld of consumption but that were later overtaken by new alternatives and that ultimately fell into disuse. Benjamin was particularly concerned with the curious way the past confronted one in these derelict places, and with the way in which such sites still contained morsels of the utopian energies that once flowed through them ([1982] 2002). In similar fashion, vaporwave frames its malls as simultaneously utopian and estranging, still conjuring dreams of consumption even as these dreams have in reality been displaced or dispelled. On vaporwave and Benjamin, see also Killeen (2018).

retention, Stiegler builds on Edmund Husserl's account of human time-consciousness and the attendant ability to apprehend temporal objects—temporal objects here referring to objects or phenomena that play out in time. Husserl frequently gives the example of the melody: a melody can only be perceived as "complete" once its final note has rung out. The object of the melody, then, is not a "static" object like a statue or a building insofar as its being is composed of a chain of moments, insofar as it exists only *as duration*. Human consciousness is nonetheless able to perceive such temporal objects as unified by synthesizing a retention of what has just passed with an apprehension of what is happening now and of what may yet come; for us to be able to perceive something as a melody, we need to be able to internally link the note we hear now with the notes that preceded it and the notes that may still arrive. Husserl supplements this "primary retention" with the notion of "secondary retention," or the capacity for memorization, anticipation, and imagination, and these two forms of retention are always in interplay: our wealth of lived experiences is sedimented as secondary retention, which impacts how we perceive the world through primary retention (my experience of a melody is, for example, affected by my familiarity with the song in question).

What Husserl, according to Stiegler, did not sufficiently recognize is that these processes of retention are always-already informed by the technical milieu in which they take place. For example, in the case of the melody, the fact that today, through a plethora of technologies, we can industrially play and replay the same song regulates the possible interplay of primary and secondary retention. The way in which a (temporal) object is technically mediated describes an irreducible part of how we experience it. This is what Stiegler calls tertiary retention: "[t]ertiary retention is in the most general sense the prosthesis of consciousness without which there could be no mind, no recall, no memory of a past that one has not personally lived, no culture" (2011a: 39). According to him, the technical carriers that mediate one's perception of the world are not inert vessels but, on the contrary, foundational to experience. This also leaves the human mind, through its technically informed capacity for primary and secondary retention, open to being reformed by the technical environment that sustains it. For Stiegler, this is an inescapable and not necessarily lamentable condition, but he worries deeply about the effects that the industrialization of temporal objects has on the human mind. This is tied to his pharmacological perspective, which posits technology as always both poison and cure—technology always augments certain perceptual or cognitive capacities while diminishing others, but for Stiegler the poisons currently far outweigh the cure.

As discussed in previous chapters, a large part of Stiegler's argument consists of a critique of how perception, memorization, and imagination have, via tertiary retention, been industrially shaped to economic ends, modified in the interest of creating circuits of short-term drives that can be temporarily mollified through consumption (2009b: 128). Tertiary retention is, to be specific, industrially molded to rouse through primary and secondary retention a perpetual drive to consume. This is achieved mainly through the widespread commercial mobilization of temporal objects; as Stiegler maintains, temporal objects are especially effective in conditioning the human mind because of how their duration coincides with the flow of consciousness (2011a: 73–4). Tellingly, the temporal objects that have for decades been drawn on to stimulate consumption are the very temporal objects that vaporwave samples: tv commercials, advertisements, jingles, muzak, commodified pop songs, video game footage, and PA announcements. These different temporal objects all coalesce within the shopping mall. Media theorist Jonathan Sterne emphasizes the extensive calculations that underlie the seemingly banal temporal objects encountered throughout the mall, amounting to a tightly controlled "production . . . of consumption" (1997: 25): advertisements, songs, and announcements are generally selected based on constantly optimized psychological insight about exactly what sounds, played at exactly what junctures, maximally stimulate what kinds of consumption (1997: 25–6). This accentuates the deep link between technology and the human mind: the strategic deployment of these industrialized temporal objects reveals human consciousness to be at least partially quantifiable and influenceable through tertiary retention and addresses the human as a malleable consumer (2009a: 41). Moreover, primary and secondary retention have grown so accustomed to these temporal objects that they now tend to largely bypass conscious reflection, disclosing a mind that has been entirely naturalized into consumerist environments.

Vaporwave's cultural resonance and its associations of nostalgia and longing disclose how inextricably entwined people have become with these technical articulations, how completely primary retention and the reservoir of secondary retentions have been shaped by temporal objects designed to smoothen consumption. The strong emotional attachment to objects and places that are entirely premised on consumption suggests that, far from encouraging a merely sterile compliance to consume, these objects and places have become part and parcel of our sense of our own past, of our innermost feelings and desires. The albums I have analyzed do not, however, simply reproduce such temporal objects in unadulterated form. Their core quality is that they refract these determinants of tertiary retention through an aesthetic of imperfection and hyper-stasis, tampering with them to shape a spectral and eerie atmosphere that distills friction from what was designed to feel frictionless. This atmosphere shows that the works of Cat System Corp., and the genre of vaporwave in general, aestheticize a more complicated relation to consumption and technology than one of carefree obedience.

Spectrality, Consumption, and Technology: Is There No Alternative?

Palm Mall's track "Veni Vidi Emi" repeats a sampled advertisement for a "summer sizzling sidewalk sale at Newmarket North Mall." In reality, this Virginia-based mall has long fallen victim to the aforementioned retail apocalypse and has been repurposed as an office center. That its devitalized echoes of habituated consumption continue to haunt the listener even as its physical referent has disappeared implies that vaporwave's imagined subjects, faced with an increasingly frictionless consumerism, continue to cling to more apprehensible temporal objects. "Veni Vidi Emi" thereby illustrates the stakes of Cat System Corp.'s hauntological aesthetic. In Chapter 2, I explained that artistic hauntology, with its aesthetic focus on imperfection and its interest in the way that technology materializes memory and anticipation, is an aesthetic means of inquiring into the compositions of tertiary retention and the (lost) pasts and futures such compositions betoken. The work of Cat System Corp. closely adheres to the sensibilities of hauntology, repurposing the ghosts of the consumerist past to dramatize a certain mode of tertiary retention that remains hauntingly active today and that curbs the human capacity to envision different technological futures.

The temporal objects that Cat System Corp. samples were originally made to radiate vitality, associated with the living, the non-ghostly: to consume is to be alive, to prosper, to stave off entropy ("I consume, therefore I am," as one of *Palm Mall's* track titles asserts). Conversely, the imperfect, hyper-static shape in which Cat System Corp.'s albums frame such vestiges of consumption implies that the mall's ghostly orbit devitalizes rather than empowers its visitors, making them seem less rather than more alive. On *NEWS AT 11's* Bandcamp page, user Calvin Yurko leaves a comment that is revealing in this light: for him, the album suggests and simultaneously unsettles the notion "of a simpler time, when McDonald's was a pleasure rather than a necessity." Indeed, according to Sean Cubitt, consumption today increasingly comprises a disciplined mode of work:

> Consumption becomes work when . . . it is undertaken not for the fulfillment of needs or the realization of aspirations, but as a disciplined function required by capital to remove the excess product manufactured

in the pursuit of expanded accumulation and growth. For capital to continue to grow, the working class of the wealthy nations now has as its chief function not mass production but the mass consumption of excess product, in cycles such as that leading from overconsumption of junk food to overconsumption of pharmaceuticals, and diet products to counter its effects. (2017a: 108)

While Cubitt unduly downplays the still salient role of mass production (for one, because the mass production of data, often realized through unrecompensed digital labor, is required to progressively refine and optimize the nets of consumption), his account is pertinent: echoing Cubitt's reading, vaporwave's spiritless assembly of temporal objects indicates a mind whose faculties of tertiary retention remain utterly conditioned by consumption. Even if this conditioning has been stripped of vigor, it remains all that vaporwave's imagined subjects know.

Cat System Corp., more pointedly, paints a world that remains haunted by the shattered "expectations of consumption inherited from the late twentieth century society, which are continuously fed by the entire apparatus of marketing and media communication" (Berardi 2011: 89). His works present consumer capitalism as a nigh inescapable condition. *Palm Mall Mars* especially charts the thorough imbrication of temporal perception, technology, and consumption; it essentially provides a technological reading of the TINA doctrine ("There Is No Alternative") and of what Mark Fisher calls capitalist realism. Fisher, indebted to Fredric Jameson, describes this latter concept as follows: "[T]he widespread sense that not only is capitalism the only viable political and economic system, but also that it is now impossible even to *imagine* a coherent alternative to it" (2009: 2, emphasis in original). *Palm Mall Mars* dramatizes the intransigence of capitalism in its consumerist manifestations. The album offers an aesthetic conception of a future whose technologies permit space travel but disallow an alternative to that pinnacle of consumerism: the mall. Notwithstanding the wealth of frictionless technologies that the year 2199 promises, the record's hauntological and hyper-static aesthetic of imperfection suggests that the faded image of the shopping mall and its familiar dreams of consumption still beset the listener, even after all these years.

In an effort to explain how capitalism has succeeded in "denying that the future is possible" and to underline hauntology's artistic value in challenging this futural foreclosure, Fisher states that "all we can expect, we have been led to believe, is more of the same—but on higher resolution screens with faster connections" (2018c: 634). This observation reveals the role the philosophy of frictionlessness plays in blocking off alternate conceptions of the future. Through preventing friction in user experience and in the navigation of commercial space, it keeps paramount the values of user-friendliness, connectivity, and optimization, and deters the emergence of any vision of an alternative. The fetishization of brighter screens that Fisher identifies, the mobilization of increasingly frictionless technologies as I have discussed it, and the apparatus of space travel as *Palm Mall Mars* imagines it all compose a mode of tertiary retention that shores up a market economy of consumption built on perpetual novelty, optimized connectivity, and short-term comfort. By integrating an image of technological expansion with an eerie and retrogressive mall-based aesthetic of imperfection, *Palm Mall Mars* suggests that the technological innovations it depicts have served only to thwart the agency to imagine alternatives to consumerism. According to Stiegler, decades of the industrial programming of tertiary retention have, indeed, induced a situation of "economic organizations taking control of the *imagination*, that is, of the *primordial source of protentions*,[15] as well as of the formation of collective secondary protentions and, ultimately, of individual and collective dreams" (2019: 47, emphasis in original). *Palm Mall Mars* takes this stunted technological imagination and projects it into a futural space, spelling out across the width of the cosmos that, truly, There Is No Alternative.

As noted earlier, *Palm Mall Mars*'s vision of the future does not directly reflect reality: mall-based forms of consumption are in retrograde across the West and have increasingly been supplanted by frictionless forms of online consumption. What does it say, then, that this album and the broader genre of vaporwave materialized precisely at a time when frictionless commerce was blooming? I argue that there are two particular conditions of frictionlessness that *Palm Mall Mars* aesthetically underlines. First, it responds to how the consumption that is today conducted online and that is encouraged through fine-grained, personalized ads provides a less collective and communal experience than the mall does and is increasingly achieved through algorithmic processes that proceed beyond the user's purview. Consumption is now largely premised on microtemporal objects that are not immediately perceivable and that thus effectuate a consumerist logic that is less directly in sync with human consciousness (cf. Hansen 2015: 57). This reveals how the form of spectrality that Chapter 2 defined as technological spectrality is latently active within vaporwave. Technological spectrality concerns the capacity for technology to operate transparently or autonomously, to

---

[15] "Protention" is another term that Stiegler borrows from Husserl. It refers to the ability to anticipate and imagine the future, which is a necessary element of time-consciousness.

circumvent the grasp of human cognition and perception.[16] This delineates not simply the increasingly transparent and frictionless devices that populate the market, such as Amazon's Echo products (a far cry from the outmoded devices vaporwave's advertisements generally endorse), but also, and primarily, the personalized and intangible processes that increasingly steer consumption. *Palm Mall Mars*, by contrast, harkens back to a time when consumerism was premised on more graspable and re-appropriable forms of tertiary retention, revealing the shared, emotional potencies such forms still harbor in an age of frictionless design.

Second, what *Palm Mall Mars*'s hyper-static aesthetic conveys is that, behind today's increasingly frictionless means of consumption and transmission, one can still hear play the archaic muzak of the mall. While frictionless technology may expand, develop, and accelerate centuries into the future, the core logic behind it remains unfailingly that of the shopping mall. Even if this site physically disappears, *Palm Mall Mars* suggests it to remain hauntingly present, still routing today's increasingly spectral devices, still urging people to keep consuming. Part of the eeriness that is proper to vaporwave thus consists in the feeling that frictionless technologies have not so much inaugurated a more frictionless or stress-free existence as they have further accelerated a consumerist logic of unceasing novelty and optimization. *Palm Mall Mars* communicates how, under present conditions, new technologies will remain haunted by the ghost of the mall and its lost futures of untroubled prosperity, making plain how a rhythm of hyper-stasis defines the philosophy of frictionlessness.

New Technological Imaginaries

Regardless of its critical ambiguity, vaporwave's practice reflects a strategy that, returning to the work of Stiegler, could help to reclaim the industrially hampered ability to conceive of alternate futures. While Stiegler does not engage the notion of hauntological aesthetics, he implicitly appeals to its capacity to bring out alternate technological potentialities when he urges

---

[16] Because of a mutual emphasis on the progressive technological decentering of the human, Grafton Tanner views vaporwave as spiritually kindred to the field of object-oriented ontology (2016: 18)—the school of thought that criticizes the standing privilege of human consciousness within philosophy and that proposes a flat ontology untainted by human subjectivity. I would not go so far as this, because no matter how uncanny vaporwave's aesthetic might seem and no matter how stripped of agency its subjects appear, the unbridled focus on consumption and the appeal to embodied memorization always reinscribe the human into the heart of the scene.

his readers to "cultivate buried anamnesic possibilities" by taking active control of the temporal objects designed to modify the mind (2019: 103). A truly liberated user must, he insists, "expect nothing from 'technological solutionism'. The moral being of the digital culture to come will be a practitioner, not a consumer" (Stiegler 2019: 295). Stiegler thus heeds against proceeding on the decade-long path of ever-optimized temporal objects that primarily reinforce a circular drive to consume, and instead maintains we should technologically reclaim the ability to determine our own needs and desires. For him, this starts with technological interventions that emphasize humanity's technological agency. Vaporwave arguably embodies one such intervention as it provides a proactive way of participating in aesthetic praxis, and of doing so largely outside the meshes of the market. It is not only that vaporwave is easy to produce and that the tools to create it are readily available online (often for free through illegal torrents), but also that its platforms of distribution, consumption, and discussion are largely located beyond the restraints of capitalist accumulation—its preferred platform of distribution Bandcamp, for instance, grants individuals the opportunity to pay what they want practically directly to the producer. Vaporwave's potentially most radical gesture is, however, its aesthetic reclaiming of consumerism's temporal objects. The albums I have analyzed all actively reconfigure signs intended to stimulate consumption, but *Palm Mall Mars*'s projection of an eternal mall especially draws attention to the thrall commercialized temporal objects have long had and continue to have over the mind, constitutive as they have become of tertiary retention.

Like the hauntological artists and objects I discussed in previous chapters, vaporwave arguably fosters a chronolibidinal attachment to technology, exploiting the links between loss and desire by rendering its objects faded and fragile. Yet, where my previously discussed objects elicited an affirmative and care-informed response, vaporwave's faded objects spell a more ambiguous channeling of chronolibido. The cultural attachment to consumerist objects that the genre elicits and explores ultimately raises questions about the pharmacologically beneficial nature of the emotionally charged relation between consciousness and consumption.

While the existential primacy of imperfection and the concepts of autoimmunity and pharmacology have so far remained unacknowledged in this chapter, they are, in fact, at the heart of questions about how to consume. Autoimmunity ensures that finitude and fragility are foundational conditions and that consumption can therefore never proceed unbounded. I have suggested that any mode of pharmacological thinking—recognizing that aesthetics play a crucial role in reinforcing or tempering the cures and toxicities of technology—must ask how users/consumers are attuned to this predicament. The forms of tertiary retention that vaporwave aesthetically reimagines and the accelerating consumerist logic behind the philosophy of frictionlessness aim to shape a libidinal economy that prevents reflections on the limits of consumption, on its value as a practice on which to build one's finite existence, and on its destructive global effects (cf. Syse and Mueller 2015). This economy's temporal objects do not encourage a technologically and chronolibidinally informed reflection on finitude and fragility that would reveal a collective responsibility of sustainable expenditure, but rather modulate a mind that perpetually desires to consume, with little eye for its spiritual and environmental implications. The forms of care discussed in previous chapters are hampered by the techno-culture that vaporwave charts, oriented as this culture is toward directing desire to novel objects and devices in order to keep extracting profit. Vaporwave, through its hyper-static aesthetic of imperfection, offers a dramatized perspective on this situation. Moreover, by finding *différance* and friction in the ghostly recurrence of the familiar, it reveals that the constituents of tertiary retention can always be reclaimed and repurposed.

There are certainly limits to vaporwave's critical potential. While I hope to have shown that his own work holds resources for tempering this skepticism, Fisher, for one, thought that vaporwave's exclusive reliance on outmoded imagery thwarted the genre's political and subversive capacities (2018a: 685). Indeed, vaporwave's wistful collage of superseded technologies does not directly reflect the technological tendencies of the contemporary moment. As technologies increasingly recede from view, making consumption progressively frictionless, vaporwave relies on a model of hauntological art that is fundamentally invested in media whose temporality and substance are directly apprehensible by human consciousness. Vaporwave, in short, hinges on and is thus entwined with the perceptibility of technological materiality. As I have insisted, the pertinence of this reliance is that it reveals a tenacious and haunting link to a certain consumption-driven composition of tertiary retention, even as technologies accelerate and themselves become more spectral. At the same time, Fisher's hesitance does suggest that vaporwave registers a situation but does not necessarily show a way out, at least not one that explicitly mobilizes today's spectral technologies. A related limit concerns vaporwave's own ambivalent stance toward its dreamworlds of consumption: some producers unquestionably criticize the collusion of consumption and technology, whereas others merely seek to bask in the glow of the 1990s shopping mall. Vaporwave's aesthetic of imperfection registers a sense of estrangement, but to what extent does it (aim to) illuminate a passage out of the deadlocks that inform its rhythm of hyper-stasis? (One might, however, argue that this ambivalence fits the genre's ghostly atmosphere.

Tying it down to one single course of action would, in fact, fail to do justice to the thoroughly ambiguous nature of the specter itself.)

Moreover, for all the unwarranted proclamations of its "completely globalized" nature (Wolfenstein osX 2015), vaporwave's virtual fantasies cannot be uncoupled from localized political, sociocultural, and economic realities—not only because vaporwave appeals to the thoroughly material and always localized technical nature of remembrance and cognition, but also because it mostly speaks to specific cultural and socioeconomic positions. The genre addresses more the experience of the Western, suburban family making their weekend trip to the mall and less the experience of the undocumented laborer who gets up at dawn to swipe floors that have been desecrated by another day of cheerful consumption. In mallsoft, the mall is framed only as a place of commercial transaction and not as one requiring constant maintenance and exploitation, making it a genre invoking mostly a particular class of people. This attests to another critical limit to vaporwave: because the clean, frictionless techno-world it repurposes glosses over the material conditions that guide the production of its commodities, vaporwave itself also has little to say about the vast networks of destruction and expropriation that define the commodity chain. The genre enacts the mental malaise such commodities may produce once populating the shelves of the mall but does not, for example, recognize the death and suffering that have predated their glistening surfaces, hiding the extortionate working conditions in global sweatshops neatly from view. Vaporwave thus highlights only one pharmacological aspect and should ideally be supplemented by other perspectives and practices if it is to generate a truly communal pharmacological vista.

Without wanting to minimize these limits, I suggest that vaporwave's technological aesthetic of imperfection enacts and makes palpable a relation to technology that is pervasive today. Stiegler argues that, even under conditions of economic turmoil and ecological catastrophe, marketeers seek to perpetuate a "bad sleep" that is awash with infinite commodities and advertisements (2019: 264). Vaporwave aestheticizes this situation from the consumer's perspective, injecting hints of friction into the dreamworlds of consumption that remain spectrally present. It reveals how technology has affected the very composition of consciousness, memory, and anticipation, and depicts how technologies have, in the past decades, been mustered to induce a cyclical short-term consumerism. In its least critical manifestations, vaporwave affectively discloses how difficult it has become to dream of a realm beyond boundless consumption. In its most critical capacities, it may open up new technological imaginaries that diverge from the consumerist logic of frictionlessness, a vital prospect for building worlds that are less driven by profit and exploitation.

Coda

On Technological Melancholia

In *Athens, Still Remains* (first published in 1996), a meditation on the work of French photographer Jean-François Bonhomme, Jacques Derrida moves from a consideration of the ontological inevitability of death to an empirical contemplation of obsolescence. *Athens, Still Remains* exhibits, in typically Derridean fashion, a sensitivity to the unassailable condition of autoimmunity—it is haunted by the recurring sentence that "we owe ourselves to death"—but is at the same time driven by an interest in finitude's material manifestations. In Bonhomme's photographs, Derrida sees a conflation of multiple temporalities of ruin and decay, all regimented by the autoimmune debt to death, yet unique in their material specificity. In their depiction of Athens as a city of ruins—some centuries old, some only seconds—these photographs encourage in Derrida "an affection for what has fallen into disuse" (2010: 47). The images comprise small inquiries into the many different forms of death that attend a general condition of finitude and convey a desire to give testimony to objects on which dust is slowly settling, gradually muted by the bustle of urban life. Bonhomme, so Derrida proposes, traces the city's marginalized objects, decrepit structures, and defunct technologies "[i]n the service not of a personal nostalgia but of a melancholy that marks a certain essence or historical experience or, if you prefer, the meaning or sense for history" (2010: 39).

Derrida's words recall Roland Barthes's famous description of "the melancholy of Photography itself" (Barthes 1981: 79). For Barthes, this melancholy derives primarily from how, in every photograph, we are confronted with "that rather terrible thing: . . . the return of the dead" (1981: 9). Indeed, the power of the camera allows each moment to live on as a trace, perhaps even to survive its referent, and a photograph by definition brings its beholder face to face with the ghost of the past (cf. Derrida 2007b: 285). Yet, while Barthes is adamant that photography is a singularly melancholic medium—"cinema," for example, is for him "in no way melancholic," as its propulsive flow prevents any image from attaching itself to the viewer (1981: 90)—I propose that the notion of melancholia has a broader value in relation to technology. In this Coda, I argue that melancholia describes a temperament, technologically elicitable by more than photography alone, that befits the current pharmacological situation. Melancholia is intimately linked to imperfection as both an existential condition and an aesthetic category, and is deeply mired in the concepts that have been so central to this study: autoimmunity, chronolibido, and spectrality. More exactly, melancholia describes a disposition that clings to rather than rejects the ghost. Following my analysis of frictionlessness as a ghost-producing design philosophy and the spectral valences of *GlitchHiker*, *The Collapse of PAL* and *Palm Mall Mars*, I contend that melancholia offers an appropriate angle from which to regard today's poisonous technological tendencies.

Melancholia

Within the long tradition of writing about melancholia,[1] the agencies and abilities that have been attributed to the melancholic subject vary greatly but almost always connote a certain relation to the spectral. We know from Sigmund Freud's seminal essay "Mourning and Melancholia" that melancholia is often taken to describe a debilitating pathology: whereas mourning marks the painful yet productive work of dissolving one's attachment to a lost object so that new attachments may be formed, the melancholic sufferer refuses to admit the actuality of loss and instead turns inward, holding on to her ghosts (1957: 249). Such a forlorn interpretation of melancholia informs, for example, political theorist Wendy Brown's idea of "left melancholia"—a term she borrows from Walter Benjamin, who was himself a notably melancholic figure (Sontag 1981: 111). Left melancholia describes a pervasive political atmosphere that demarcates a

> Left that has become more attached to its impossibility than to its potential fruitfulness, a Left that is most at home dwelling not in hopefulness but in its own marginality and failure, a Left that is thus caught in a structure of melancholic attachment to a certain strain of its own dead past, whose spirit is ghostly, whose structure of desire is backward looking and punishing. (Brown 2003: 464)

Brown identifies a melancholia that is inward-looking and bereft of hope; that, for all it seems, has abandoned the future, crestfallenly shuffling amongst the ghosts of its losses.

---

[1] For further reading, see, in addition to the titles mentioned in this coda, Klibansky, Panofsky, and Saxl ([1964] 2019); Bowring (2015).

In *Ghost of my Life* (2014), Mark Fisher contrasts Brown's left melancholia with a more politically potent and future-oriented form of melancholia that is nonetheless still steeped in the spectral: hauntological melancholia. For Fisher, this melancholia is proper to the work of hauntologically inclined artists like Burial and William Basinski. In previous chapters, I have discussed the notion of hauntological art and have stressed that its aesthetic features always tend to contain, no matter how remote, the promise of an alternate future. The hauntological melancholia that seeps from such works indeed discloses a "refusal to give up on the desire for the future. This refusal gives the melancholia a political dimension, because it amounts to a failure to accommodate to the closed horizons of capitalist realism" (Fisher 2014: 21). Here, the melancholic is certainly aware of the many erasures gathered under the *no longer* of her ghosts, but this does not render her entirely dejected; she recognizes that the structure of spectrality always holds out the *not yet* of a different future. The unwillingness to give up on the lost object becomes a politically charged "refusal to adjust to what current conditions call 'reality'" (Fisher 2014: 24). This more affirmative take on melancholia resonates with the way melancholia is figured in the edited volume *Loss* (2003). In this influential volume, critical theorists David L. Eng and David Kazanjian underline melancholia's open-ended, positive, and creative qualities. The melancholic temperament, they suggest, "offers a capaciousness of meaning in relation to losses encompassing the individual and the collective, the spiritual and the material, the psychic and the social, the aesthetic and the political" (Eng and Kazanjian 2003: 3). In this interpretation, melancholia is hailed as empowering because it makes one decline to move on; instead, it establishes "an ongoing and open relationship with the past—bringing its ghosts and specters, its flaring and fleeting images, into the present" (Eng and Kazanjian 2003: 4).[2]

These disparate accounts are united by an insistent motif: melancholia is figurative of a relation to loss (of a person, an object, a futural trajectory) that does not seek to do away with the ghostly remains (which is how mourning, as the uncomfortable but progressive work of dissipating attachments, is often theorized) but rather holds on to them. This attachment may take many different forms—from a pathological refusal to accept the object's ghostly

---

[2] Literary scholar Eugenie Brinkema has criticized *Loss* as emblematic of the trend to frame melancholia predominantly, or even exclusively, as a capacitating experience of loss: "In this celebration of the formerly pathologized term, . . . the problems of illumination, mysteriousness, and affective pain . . . are set aside" (2014: 65–6); melancholia thereby threatens to become "all *Schein* and no *Schmerz*" (2014: 65, emphasis in original). Indeed, it is important not to theorize melancholia too readily as only an affirmative sensibility, lest one neglect its always-latent risk of inducing depression, immobilization, and ennui.

aspects to the political struggle for which Fisher hopes. The point, however, is that melancholia, in its numerous manifestations, describes a temperament or feeling that is tied in a distinct way to spectrality, finitude, and loss: while it is rife with ambivalences, it always transmits the unmistakable sense of attuning oneself to loss in a way that does not aim to exorcize the ghost.

It is for this reason that the melancholic disposition appears to be most fitting to describe both the ethical and political implications of the human perception of the existential primacy of imperfection *and* to find an appropriate response to the pharmacological situation today's frictionless technologies induce. Cultural theorist László Földényi's comprehensive 1984 study *Melancholy*, which elucidates this often ill-defined concept, helps to further unravel the first part of my claim. Földényi traces melancholia's history, extending far beyond the strictly pathological character that Freud ascribed to it, and shows how it has always been a term associated with mortality, ambiguity, and imperfection. As Földényi suggests, melancholia, more than related notions like depression and nostalgia, connotes a deep insight "into the interconnections of existence" ([1984] 2016: 252). More specifically, I argue that melancholia, as a temperament imprinted with temporality and loss, offers a palpable experience of the existential primacy of imperfection and of the conditions of autoimmunity, chronolibido, and spectrality that have guided this study.

The preliminary definition of melancholia that Földényi gives is worth citing at length:

> Melancholia is, among other things, a consequence of the inadequacy of concepts; that inadequacy, however, is not some kind of deficiency that can be overcome or even eliminated over time, but the sort of thing without which concept formation is unimaginable, and just as clear-sightedness, measure, or definitiveness forms one of the pillars on which all insight can rest, so obscurity, gloom, incomprehensibility, and dissatisfaction form the other. Hence, perhaps, the sadness that lurks in the depths of any formulation laying claim to finality, the inconsolability that corrodes even the most closed formations. ([1984] 2016: 4)

Földényi further underlines his claim that melancholia is an effect of the necessity of contingency when he writes that "[m]elancholia does not emerge from sheer order (which does not exist anyway) but out of the inevitable cracks concealed in order" ([1984] 2016: 306). There is a Derridean spirit to Földényi's work, owing to its persistent appeal to the impossibility of purity and wholeness. Indeed, the observation that there cannot but be cracks within the bastion of order recalls the condition of autoimmunity—

the constitutive condition, stemming from the spacing of time, of always-already being threatened from within, of already containing within oneself an irreducible opening to the specter of finitude and violence. Földényi locates the cause of melancholia precisely in the apprehension of this fundamental inadequacy. The concept of autoimmunity becomes even more tangible when Földényi describes those thinkers that prove insusceptible to the melancholic temperament: "Thinkers who apprehend the world in its prosaic one-sidedness naturally reject the possibility of identifying life with death, suffering with enjoyment. Nor can they otherwise, since they see the world as a multiplicity of incomprehensible and senseless things in which they *cannot discover the inner kinship of contradictory phenomena,* only their mutually excluding antagonism" ([1984] 2016: 215, emphasis added). Autoimmunity registers exactly this "inner kinship of contradictory phenomena": it discloses that life cannot but be threatened by death, memory cannot but be threatened by forgetting, perfection cannot but be threatened by imperfection. Melancholics, Földéniy's work divulges, perceive that the corruptive threat of alterity that autoimmunity brings suffuses existence.

Melancholics, to phrase this differently, intuit the existential primacy of imperfection. They sense that, if there is a logic that permeates existence, it is not a progressive perfectibility or celestial harmony, but rather an indelible finitude that unconditionally haunts all things. Tellingly, Földényi describes the modern melancholic, in contrast to the medieval thinker who grieved the "distance from divine perfection," as a figure that is "melancholic not because of a *yearning* for perfection but because of the *reality* of the imperfection that exists" ([1984] 2016: 186, emphasis in original). Melancholia thus stems primarily from the awareness that there is no perfect design to guide existence; rather, it is imperfection that is tantamount, making the risk of loss unassailable.

Martin Hägglund's notion of chronolibido, a concept that has been central to my aesthetic analyses, reveals this risk to ground all care and desire: only an object that is finite and therefore losable can elicit a libidinal investment (2012: 9). The melancholic proves herself keenly aware of this situation: she feels that even, or especially, seemingly "perfect works [are] always haunted by the possibility of annihilation," an annihilation that is not symbolic but rather symptomatic of the very real death "that has to be confronted by every solitary person" (Földényi [1984] 2016: 140). For the melancholic, the prospects of loss and finitude may produce an excess of both chronophobia (a fear of time) and chronophilia (a love of time), but they first of all generate a sensitivity to "the incessant changeability of existence" (Földényi [1984] 2016: 297), to a "situation's exceptionality and its being bound to the moment" (Földényi [1984] 2016: 133). Földényi unwittingly aligns himself with Hägglund and Derrida when he argues that this changeability and the sufferings and losses it may yield are inextricably bound to the joys of existence: "[S]uffering does not arrive from outside but is latent in man, ready to burst out at any moment. Therefore, to deprive man of his suffering, however being the hope, is to pronounce a death sentence (only in death is there no suffering)" ([1984] 2016: 286). This is not to say that the melancholic *desires* loss or suffering—Földényi describes, for example, how the melancholic may come to experience the future as a wellspring of agony because it signals the evanescence of the present ([1984] 2016: 315). The point, however, is that more than other temperaments, melancholia is not simply governed by but actively attunes one to the centrality of loss to existence. As Susan Sontag claims, "[p]recisely because the melancholy character is haunted by death, it is melancholics who best know how to read the world" (1981: 119–20).

This distinct relation to loss returns us once more to the figure of the ghost. As suggested, melancholia connotes a refusal to lay ghosts to rest; it does away with mourning's naive suggestion that wounds can cleanly heal and that ghosts can tidily depart. As literary scholars Klaus Mladek and George Edmondson contend in their call for a politics of melancholia, dreams of a future "free and unmarred by specters of the more disturbing kind, may appeal to the nostalgic and the mournful, but they are never for the melancholic" (2009: 230). Rather, "melancholics are those who invite the ghosts to the table, who welcome their arrival, who affirm their intrusion" (Mladek and Edmondson 2009: 230). Here, we find a crucial difference between melancholia and nostalgia: while melancholia may always slide into nostalgia (does Brown's left melancholia not also describe a kind of left nostalgia?), there is also something in melancholia that exceeds the nostalgist's stagnant attachment to past objects. While the nostalgist stands transfixed—often reactionarily so—by a calcified image of the past, the melancholic allows the specter more room to assert itself. In inviting ghosts to her table, the melancholic does not, or at least not necessarily, desire to fixate them, but rather aims for a more open-ended relation. Echoing Esther Peeren's conceptualization of spectrality as a perspective that fosters attention for the marginalized (2014: 13), Mladek and Edmondson assert that "the melancholic assumes the burden of what we carry on our back; he counts with what is not counted, what remains unnamed and drops out of symbolic representation" (2009: 215). The melancholic, in other words, is responsive to the ghost's proclivity to testify to the erased and the unheeded.

Melancholia, then, does not merely entail an admittance of the ghost, but also an (often debilitating) awareness of its contingent aspects. Because of the melancholic's attunement to the mutable and fragile elements of existence, she is painfully aware that the ghostly sufferings and miseries she encounters are not necessary sufferings and miseries. They are, rather, the result of material, historical processes that might have been prevented or rerouted and that may implicate her. The melancholic subject feels she may be answerable and responsible for how a general condition of finitude and imperfection has been narrowed down to produce this particular ghostly ensemble. Yet, this sensitivity to the contingent also tells her that things can be different; if, as Földényi insists, the melancholic reads in loss a sign of the "incessant changeability of existence" ([1984] 2016: 297), she also recognizes that past and present injustices may be atoned for in the future. Invoking the injustices produced under today's political and judicial "machinery of suffering and salvation," Mladek and Edmondson ask whether it is not "surprising that the more we service this machinery, the more ghosts we produce" (2009: 223). "Perhaps," they proceed, "we have invented an apparatus around the wrong set of questions and answers. Perhaps we need to raise the possibility that the anxiety ghosts provoke points to our complicity in a more fundamental betrayal" (2009: 223). Can a similar question not be asked in relation to the philosophy of frictionlessness and its pharmacological composition?

As I have demonstrated, frictionlessness functions as a ghost-producing machinery that aesthetically perpetuates itself by emptying perception of signs of toil while expanding an underlying system of extraction and exploitation. Pharmacologically speaking, one of the most toxic effects of frictionlessness is that it hides so much of its poison from the gaze of the user, that so much of its ghosts are ghostly because they are kept from scrutiny and thereby from the capacity to solicit care. Might melancholia, as an attitude that actively engages the spectral, offer a more hospitable look for these ghosts? It is telling that media theorist Sean Cubitt has recourse to the concept of melancholia to specify the burden bestowed upon us humans in the face of technology's destructive effects. He describes how the melancholic gaze, as it dwells over barren landscapes wrought by the manufacture and disposal of supposedly unsullied devices, proves cognizant of "the lost, the silenced, and the unsuccessfully erased, human and nonhuman alike," creating an opening to bear a duty to these spectral shapes (Cubitt 2017a: 187). Could we, however, also conceive of a technologically modulated melancholia that does not require the palpable despoliation of the planet to become invested? Could we conceive of a melancholic, pharmacologically curative engagement of technology that prevents the ongoing intensification of destruction and exploitation? Such a melancholia, affected by the spectral, would begin with seeking out a different "set of questions and answers" (Mladek and Edmondson 2009: 223) around which to design technologies. As I have asked through the objects analyzed in Chapters 3 to 5, what would it mean if we were to aesthetically reimagine technology by starting with questions of care, finitude and fragility—what would it mean if technology would make us more tangibly acknowledge the existential primacy of imperfection?

Technological Melancholia

The objects I have examined all draw on an aesthetic of imperfection to shape a relation to technological finitude that is tinged with melancholia. In *GlitchHiker*, the positive charge of playing the game was infused with the negative charge of absence and loss; the players' most defining emotions were empathy and guilt over contributing to the game's extinction. *The Collapse of PAL*, while at face value perhaps appearing to stage a work of mourning, is, in fact, more a work of melancholia; at no point does the performance imply that PAL's ghost should be successfully grieved so that we can, with clear conscience, move on to more frictionless alternatives. Rather, Menkman employs a melancholic gaze that looks back on the specters of the technological past with care and concern, discovering new ways of making them address the present. The work of vaporwave producer Cat System Corp. dramatizes a mind that cannot rid itself of the technological, consumer-oriented objects of its past, but it challenges a simple nostalgia by imbuing its sounds with hints of friction. The dead or dreamlike malls Cat System Corp.'s records depict reveal how tightly technology, consumption, and consciousness have become interwoven in consumerist societies, but their reconfiguration of the familiar also attests to the possibility of transformation. While, for Barthes, the feeling of photographic melancholy sprung mainly from the pastness of the referent, which renders the unyielding ravages of time sensible, the preceding examples show that there are other medial, aesthetic properties that can invoke a sense of melancholia. *Technological melancholia*, then, may be defined as the experiences of melancholy that attach to and are shaped by the technological calibration and aesthetic presentation of the temporal.[3] The objects I have analyzed establish forms of "tertiary retention"—to borrow Bernard Stiegler's term for the exteriorized, technical elements of temporal consciousness—that place finitude and fragility at the center of perception, encouraging an alternate, more melancholically infused attachment to technology. These haunted artworks thereby show that an aesthetic reclamation of the technical

---

[3] Contrasting interpretations of melancholia in a technological context exist, such as media scholar Geert Lovink's work on sadness and melancholia as technological design functions, describing a sort of "techno-sadness" that comes with being "caught in the perpetual now" (55).

prostheses of consciousness may open up new dispositions, new futural forms, new practices of care for the technological realm, new ensembles of technological ghosts.

In the Introduction, I drew on philosopher Yuriko Saito's observation of the "considerable power of the aesthetic to guide people's behavior, decisions, and actions" (2017a: 141). I also cited environmental scholars Robert S. Emmett and David E. Nye's argument about knowledge needing to be "*affective*, or emotionally potent, in order to be *effective*, or capable of mobilizing social adaptation" (2017: 8, emphasis in original). The hauntological aesthetic of imperfection that marks my analyzed objects offers empirical proof of these suggestions, showing how different aesthetic articulations of tertiary retention allow for new conceptions of and relations to technology and its finitudes. These objects reveal that there are many possible questions around which we can design technological devices—questions that might thwart today's destructive norms of frictionlessness. What if we pay more attention to the autoimmunity of technology and remain mindful of the material effects of the constitutive condition of imperfection? What if we think with the concept of chronolibido and envision a politics of technology that more consciously integrates the co-implication of loss and desire into its designs? What if, in contrast to the spectralizing machinery of frictionlessness, we pay more heed to the ghostly when inventing our devices? What if we elevate care to a guiding principle and refuse to let the value of optimization render our relation to individual devices transient? My study has shown that an ethics of pharmacological thinking emerges from these questions, whereby, always positing technology as both poison and cure, one remains responsible for whatever toxic effects spill from technological assemblages and their ancillary modes of tertiary retention. A sentiment of melancholia, responsive to rather than dismissive of the ghosts that past and current conditions have bred, appears to be an appropriate disposition for addressing such questions. Always vacillating between hope and despondency, melancholics sense that the path to a more hospitable future must be trod in the company of the lost and the dead (Cubitt 2017a: 192). An aesthetic of imperfection and the friction it produces prove conducive to the rise of melancholia, clouding one's perception of the technological world with the pangs and pulls of finitude and fragility. What exact form melancholia takes is never guaranteed—and it is crucial to stress that it always comes with risks, such as immobilization and the problematic romanticization of pain—but finding different affective and perceptual involvements with technology constitutes an important step on the long road toward more collectively curative pharmacologies.

Derrida saw, emanating from Bonhomme's photographs, a ghostly writing that read, "we owe ourselves to death." I have argued that a truly sustainable pharmacology would produce technologies that tell us that we owe ourselves to the deaths of others as well. On an existential level, this involves the recognition of finitude and fragility as unassailable and collective conditions. On a material level, this entails the awareness that there are always ghosts in our machines, even if contemporary technologies try violently to keep them from view. A technological aesthetic of imperfection can bind these two dimensions and, in its most critical capacity, confers a ghostly and melancholic burden that must be borne with care, empathy, and concern.

Bibliography

Adams, T. (2013), "Google and the Future of Search: Amit Singhal and the Knowledge Graph," *The Guardian*, January 19. Available online: https://www .theguardian.com/technology/2013/jan/19/google-search-knowledge-graph -singhal-interview (accessed March 27, 2023).

Agamben, G. (2013), "On the Uses and Disadvantages of Living among Specters," in M. Blanco and E. Peeren (eds.), *The Spectralities Reader: Ghosts and Haunting in Contemporary Cultural Theory*, 473–7, New York: Bloomsbury.

Ahmed, S. (2010), *The Promise of Happiness*, Durham and London: Duke University Press.

Andrews, I. (2002), "Post-Digital Aesthetics and the Return to Modernism," *ian -andrews.org*. Available online: www.ian-andrews.org/texts/postdig.html (accessed October 31, 2022).

Antonelli, P. (2010), "Hella's Imperfect World," in H. Jongerius (ed.), *Misfit*, 233–4, London: Phaidon Press Limited.

Appiah, K. A. (2017), *As If: Idealization and Ideals*, Cambridge, MA and London: Harvard University Press.

Arboleda, M. (2020), *Planetary Mine: Territories of Extraction under Late Capitalism*, London and New York: Verso.

Arcibal, C. (2019), "China's Outlet Mall Boom is Proving Immune to the Online Shopping Threat, and More Investment Funds are Taking a Look," *South China Morning Post*, July 11. Available online: www.scmp.com/property /hong-kong-china/article/3018095/chinas-outlet-mall-boom-proving -immune-online-shopping (accessed October 31, 2022).

Ashbaugh, D., W. Gibson, and K. Begos (1992), *Agrippa (A Book of the Dead)*, New York: Kevin Begos.

Augé, M. ([1992] 2009), *Non-Places: Introduction to an Anthropology of Supermodernity*, trans. J. Howe, London and New York: Verso.

Babich, N. (2018), "Frictionless Experience: How to Create Smooth User Flows," *Shopify*, July 3. Available online: https://www.shopify.com/partners/blog/user -flow (accessed October 31, 2022).

Bal, M. (1996), *Double Exposures: The Subject of Cultural Analysis*, New York and London: Routledge.

Baraniuk, C. (2013), "Glitchland: In the Future, the Digital Will Know How to Decay," *The Machine Starts*, September 14. Available online: web.archive .org/web/20180814104606/http://www.themachinestarts.com/read/2013-09 -glitchland-future-digital-will-know-how-to-decay (accessed October 31, 2022).

Barrow Jr., M. V. (2009), *Nature's Ghosts: Confronting Extinction from the Age of Jefferson to the Age of Ecology*, Chicago and London: The University of Chicago Press.

Barthes, R. (1981), *Camera Lucida: Reflections on Photography*, trans. R. Howard, New York: Hill and Wang.

Basinski, W. (2002–3), *The Disintegration Loops I-IV*, 2062.

Beller, J. (2018), *The Message is Murder: Substrates of Computational Capital*, London: Pluto Press.

Benjamin, W. ([1982] 2002), *The Arcades Project*, trans. H. Eiland and K. McLaughlin, Cambridge and London: The Belknap Press of Harvard University Press.

Benjamin, W. ([1928] 2009), *The Origin of German Tragic Drama*, trans. J. Osborne, London and New York: Verso.

Benjamin, W. ([1955] 2015), "Theses on the Philosophy of History," in H. Arendt (ed.), *Illuminations*, trans. H. Zorn, 245–55, London: The Bodley Head.

Bennington, G. and J. Derrida (1993), *Jacques Derrida*, trans. Geoffrey Bennington, Chicago and London: The University of Chicago Press.

Berardi, F. B. (2011), *After the Future*, eds. G. Genosko and N. Thoburn, Oakland and Edinburgh: AK Press.

Berardi, F. B. (2015), *And: Phenomenology of the End*, Los Angeles: Semiotext(e).

Berlant, L. (2011), *Cruel Optimism*, Durham and London: Duke University Press.

Berry, D. M. and M. Dieter, eds. (2015a), *Postdigital Aesthetics: Art, Computation and Design*, New York: Palgrave Macmillan.

Berry, D. M. and M. Dieter (2015b), "Thinking Postdigital Aesthetics: Art, Computation and Design", in D. M. Berry and M. Dieter (eds.), *Postdigital Aesthetics: Art, Computation and Design*, 1–11, New York: Palgrave Macmillan.

Berry, D. M. and A. Fagerjord (2017), *Digital Humanities*, Cambridge: Polity Press.

Betancourt, M. (2017), *Glitch Art in Theory and Practice: Critical Failures and Post-Digital Aesthetics*, New York and London: Routledge.

Bishop, R., K. Gansing, J. Parikka, and E. Wilk, eds. (2016), *Across & Beyond: A Transmediale Reader on Post-Digital Practices, Concepts, and Institutions*, Berlin: Sternberg Press and Transmediale e.V.

Blanco, M. and E. Peeren (2013), "The Ghost in the Machine: Spectral Media/ Introduction," in M. Blanco and E. Peeren (eds.), *The Spectralities Reader: Ghosts and Haunting in Contemporary Cultural Theory*, 199–206, New York: Bloomsbury.

Boltanski, L. and E. Chiapello (2007), *The New Spirit of Capitalism*, trans. Gregory Elliott, London and New York: Verso.

Bonnet, F. J. (2020), *After Death*, trans. A. Ireland and R. Mackay, Falmouth: Urbanomic.

Born, G. and C. Haworth (2017), "From Microsound to Vaporwave: Internet-mediated Musics, Online Methods and Genre," *Music and Letters*, 98 (4): 601–47.

Borradori, G. (2003), *Philosophy in a Time of Terror: Dialogues with Jürgen Habermas and Jacques Derrida*, Chicago and London: The University of Chicago Press.

Bowe, M. (2013), "Q&A: James Ferraro On NYC's Hidden Darkness, Musical Sincerity, And Being Called the God of Vaporwave," *Stereogum*, October 11. Available online: www.stereogum.com/1504091/qa-james-ferraro-on-nycs-hidden-darkness-musical-sincerity-and-being-called-the-god-of-vaporwave/interviews/ (accessed October 31, 2022).

Bowring, J. (2015), *A Field Guide to Melancholy*, Harpenden: Oldcastle Books.

Brinkema, E. (2014), *The Forms of the Affects*, Durham and London: Duke University Press.

Brock Jr., A. (2020), *Distributed Blackness: African American Cybercultures*, New York: New York University Press.

Brøvig-Hanssen R. and A. Danielsen (2016), *Digital Signatures: The Impact of Digitization on Popular Music Sound*, Cambridge, MA and London: MIT Press.

Brown, W. (2003), "Resisting Left Melancholia," in D. L. Eng and D. Kazanjian (eds.), *Loss*, 458–65, Berkeley: University of California Press.

Browne, S. (2015), *Dark Matters: On the Surveillance of Blackness*, Durham and London: Duke University Press.

Bueno, C. C. (2017), *The Attention Economy: Labour, Time and Power in Cognitive Capitalism*, London and New York: Rowman & Littlefield.

Cambridge University Press (2022), *Cambridge Dictionary*. Available online: https://dictionary.cambridge.org/ (accessed October 31, 2022).

Cascone, K. (2000), "The Aesthetics of Failure: 'Post-Digital' Tendencies in Contemporary Computer Music," *Computer Music Journal*, 24 (4): 12–18.

Cat System Corp. (2014a), 太陽。慰安。楽園.

Cat System Corp. (2014b), *Palm Mall*, Geometric Lullaby.

Cat System Corp. (2015), *Class of '84*, Colosseum.

Cat System Corp. (2016), *NEWS AT 11*.

Cat System Corp. (2018a), *A Class in... CRYPTO CURRENCY*.

Cat System Corp. (2018b), *Palm Mall Mars*.

Chandler, L. and D. Livingston (2012), "Reframing the Authentic: Photography, Mobile Technologies and the Visual Language of Digital Imperfection," *Proceedings of the 6th Global Conference, Visual Literacies: Exploring Critical Issues*, 1–15, Oxford: Inter-Disciplinary Press.

Chandler, S. (2016), "Genre as Method: The Vaporwave Family Tree, From Eccojams to Hardvapour," *Bandcamp*, November 21. Available online: daily.bandcamp.com/lists/vaporwave-genres-list (accessed October 31, 2022).

Chang, A. Y. (2019), *Playing Nature: Ecology in Video Games*, Minneapolis and London: University of Minnesota Press.

Chéroux, C. (2003), *Fautographie: Petite histoire de l'erreur photographique*, Liège: Vervinckt & fils.

Chuck Person (2010), *Chuck Person's Eccojams Vol. 1.*

Chun, W. H. K. (2011), *Programmed Visions: Software and Memory*, Cambridge, MA and London: MIT Press.

Chun, W. H. K. (2016), *Updating to Remain the Same: Habitual New Media*, Cambridge, MA and London: MIT Press.

Cloninger, C. (2011), "GltchLnguistx: The Machine in the Ghost / Static Trapped in Mouths," in N. Briz, E. Meaney, R. Menkman, W. Robertson, J. Satrom and J. Westbrook (eds.), *Glit.tc/h Reader[ror] 20111*, 23–41, Amsterdam and Chicago: Unsorted Books.

Clough, P. T. (2004), "Future Matters: Technoscience, Global Politics, and Cultural Criticism," *Social Text*, 22 (3): 1–23.

Clune, M. W. and M. Hägglund (2015), "Time in Our Time: Clune and Hägglund Debating at Stanford," *CR: The New Centennial Review*, 15 (3): 109–34.

Colvin, B. and R. Derousseau (2017), "Jeff Bezos's War with Friction," *Fortune*, February 2. Available online: https://fortune.com/2017/02/02/jeff-bezoss-war-with-friction/ (accessed March 27, 2023).

Contreras-Koterbay S. and L. Mirocha (2016), *The New Aesthetic and Art: Constellations of the Postdigital*, Amsterdam: Institute of Network Cultures.

Corsa, Z. (2012), "A Landscape of Decay," *A Closer Listen*, April 25. Available online: acloserlisten.com/2012/04/25/a-landscape-of-decay/ (accessed October 31, 2022).

Cramer, F. (2015), "What is 'Post-Digital?'" in D. M. Berry and M. Dieter (eds.), *Postdigital Aesthetics: Art, Computation and Design*, 12–26, New York: Palgrave Macmillan.

Crawford, K. (2021), *Atlas of AI: Power, Politics, and the Planetary Costs of Artificial Intelligence*, New Haven and London: Yale University Press.

Crawford, K. and V. Joler (2018), "Anatomy of an AI System: The Amazon Echo as an Anatomical Map of Human Labor, Data and Planetary Resources," *AI Now Institute and Share Lab*, September 7. Available online: anatomyof.ai/ (accessed October 31, 2022).

Cubitt, S. (2014), "Angelic Ecologies," *Millennium Film Journal*, 1 (58): 40–5.

Cubitt, S. (2017a), *Finite Media: Environmental Implications of Digital Technologies*, Durham and London: Duke University Press.

Cubitt, S. (2017b), "Glitch," *Cultural Politics*, 13 (1): 19–33.

Delfanti, A. (2021), *The Warehouse: Workers and Robots at Amazon*, London: Pluto Press.

Derrida, J. (1973), *Speech and Phenomena: And Other Essays on Husserl's Theory of Signs*, trans. D. B. Allison, Evanston: Northwestern University Press.

Derrida, J. (1978a), "Freud and the Scene of Writing," in J. Derrida, *Writing and Difference*, trans. A. Bass, 246–91, New York and London: Routledge.

Derrida, J. (1978b), "Violence and Metaphysics," in J. Derrida, *Writing and Difference*, trans. A. Bass, 97–192, New York and London: Routledge.

Derrida, J. (1981a), *Dissemination*, trans. B. Johnson, London: The Athlone Press.

Derrida, J. (1981b), *Positions*, trans. A. Bass, Chicago and London: The University of Chicago Press.

Derrida, J. (1982a), "Différance," in J. Derrida, *Margins of Philosophy*, trans. Alan Bass, 1–28, Chicago and London: The University of Chicago Press.

Derrida, J. (1982b), "The Ends of Man," in J. Derrida, *Margins of Philosophy*, trans. Alan Bass, 109–26, Chicago and London: The University of Chicago Press.

Derrida, J. (1994), *Specters of Marx: The State of the Debt, the Work of Mourning and the New International*, trans. P. Kamuf, New York and London: Routledge.

Derrida, J. (1995), *Archive Fever: A Freudian Impression*, trans. E. Prenowitz, Chicago and London: The University of Chicago Press.

Derrida, J. (2002), *Acts of Religion*, ed. G. Anidjar, New York and London: Routledge.

Derrida, J. (2005), *Paper Machine*, trans. R. Bowlby, Stanford: Stanford University Press.

Derrida, J. (2007a), "Psyche: Invention of the Other," in P. Kamuf and E. Rottenberg (eds.), *Psyche: Inventions of the Other, Volume I*, trans. C. Porter, 1–47, Stanford: Stanford University Press.

Derrida, J. (2007b), "The Deaths of Roland Barthes," in P. Kamuf and E. Rottenberg (eds.), *Psyche: Inventions of the Other, Volume I*, P. Brault and M. Naas, 264–98, Stanford: Stanford University Press.

Derrida, J. (2010), *Athens, Still Remains. The Photographs of Jean-François Bonhomme*, trans. P Brault and M. Naas, New York: Fordham University Press.

Derrida, J. and B. Stiegler (2002), *Echographies of Television*, trans. J. Bajorek, Cambridge: Polity Press.

DeSilvey, C. (2017), *Curated Decay: Heritage Beyond Saving*, Minneapolis and London: University of Minnesota Press.

Devine, K. (2019), *Decomposed: The Political Ecology of Music*, Cambridge, MA and London: MIT Press.

Dierckx, M. (2011), "De Derde Nominatie voor de Control Industry Award Is . . . GlitchHiker van Aardbever," *Control Online*, November 9. Available online: control-online.nl/gamesindustrie/2011/11/09/nieuws-de-derde-nominatie-voor-de-control-industry-award-is%E2%80%A6-glitchhiker-van-aardbever/ (accessed October 31, 2022).

D'Ignazio C. and L. F. Klein (2020), *Data Feminism*, Cambridge, MA and London: MIT Press.

DreamWorks (2018), "Album Review: 猫 シ Corp. – Palm Mall Mars," *VapourBan*, August 23. Available online: web.archive.org/web/20181215173335/https://vapourban.com/2018/08/album-review-%E7%8C%AB-%E3%82%B7-corp-palm-mall-mars/ (accessed October 31, 2022).

Duffy, M. (2017), "Doing the Dirty Work: Gender, Race and Reproductive Labor in Historical Perspective," *Gender & Society*, 12 (3): 313–36.

Dyer-Witheford, N. (2015), *Cyber-Proletariat: Global Labour in the Digital Vortex*, London: Pluto Press.

Ehrenreich, B. (2010), *Bright-Sided: How Positive Thinking is Undermining America*, London: Picador.

Elsaesser, T. (2016), *Film History as Media Archaeology*, Amsterdam: Amsterdam University Press.

Emmett, R. S. and D. E. Nye (2017), *The Environmental Humanities: A Critical Introduction*, Cambridge, MA and London: MIT Press.

Eng, D. L. and D. Kazanjian (2003), "Mourning Remains," in D. L. Eng and D. Kazanjian (eds.), *Loss*, 1–23, Berkeley: University of California Press.

Facebook IQ (2017), "Zero Friction Future," *The Economist*. Available online: https://zerofrictionfuture.economist.com/ (accessed March 27, 2023).

Fassin, D. (2009), "Another Politics of Life is Possible," *Theory, Culture & Society*, 26 (5): 44–60.

Ferraro, J. (2011), *Far Side Virtual*, Hippos in Tanks.

Fisher, M. (2009), *Capitalist Realism: Is There No Alternative?* Winchester and Washington: Zero Books.

Fisher, M. (2014), *Ghosts of My Life: Writings on Depression, Hauntology and Lost Futures*, Winchester and Washington: Zero Books.

Fisher, M. (2016), *The Weird and the Eerie*, London: Repeater Books.

Fisher, M. (2018a), "Hauntology, Nostalgia and Lost Futures: Interviewed by Valerio Mannucci and Valerio Mattioli for *Nero* (2014)," in D. Ambrose (ed.), *k-punk: The Collected and Unpublished Writings of Mark Fisher (2004–2016)*, 683–9, London: Repeater Books.

Fisher, M. (2018b), "The Outside of Everything Now," in D. Ambrose (ed.), *k-punk: The Collected and Unpublished Writings of Mark Fisher (2004–2016)*, 297–302, London: Repeater Books.

Fisher, M. (2018c), "They Can Be Different in the Future Too: Interviewed by Rowan Wilson for *Ready Steady Book* (2010)," in D. Ambrose (ed.), *k-punk: The Collected and Unpublished Writings of Mark Fisher (2004–2016)*, 627–36, London: Repeater Books.

Fisken, T. (2011), "The Spectral Proletariat: The Politics of Hauntology in *The Communist Manifesto*," *Global Discourse*, 2 (2): 17–31.

Flores, R. (2011), "GlitchHiker: The Death of a Newborn Indie Game," *Bitmob*, December 16. Available online: web.archive.org/web/20120109032626/https://bitmob.com/articles/newborn-baby-game-dies (accessed October 31, 2022).

Földényi, L. F. ([1984] 2016), *Melancholy*, trans. T. Wilkinson, New Haven and London: Yale University Press.

Ford, L. G. (2019), *Savage Messiah*, London and New York: Verso.

Foss, M. (1946), *The Idea of Perfection in the Western World*, Princeton and Oxford: Princeton University Press.

Freud, S. (1957), "Mourning and Melancholia," in J. Strachey (ed.), *The Standard Edition of the Complete Psychological Works of Sigmund Freud, Volume XIV (1914–1916)*, 237–58, London: The Hogarth Press.

Funk, T. (2018), "Dirty Your Media: Artists' Experiments in Bio-Sovereignty," in N. Lushetich (ed.), *The Aesthetics of Necropolitics*, 157–80, London and New York: Rowman & Littlefield.

Gansky, A. E. (2014), "'Ruin Porn' and the Ambivalence of Decline: Andrew Moore's Photographs of Detroit," *Photography & Culture*, 7 (2): 119–39.

Gates, B. (1995), *The Road Ahead*, New York: Viking Penguin.

Geometric Lullaby (2018), "Palm Mall Mars [Remastered]," *Bandcamp*, June 17. Available online: https://geometriclullaby.bandcamp.com/album/palm-mall -mars-remastered (accessed October 31, 2022).

Gibb, R. (2011), "Adventures On The Far Side: An Interview With James Ferraro," *The Quietus*, December 15. Available online: thequietus.com/articles /07586-james-ferraro-far-side-virtual-interview (accessed October 31, 2022).

Glitch x86 (2020), "GlitchHiker Gameplay," YouTube, February 7. Available online: www.youtube.com/watch?v=r2MDgKGCF88 (accessed October 31, 2022).

Glitsos, L. (2017), "Vaporwave, or Music Optimised for Abandoned Malls," *Popular Music*, 37 (1): 100–18.

Goodman, S., T. Heys, and E. Ikoniadou (2019), *Unsound: Undead*, Falmouth: Urbanomic.

Gordon, A. F. (1997), *Ghostly Matters: Haunting and the Sociological Imagination*, Minneapolis and London: University of Minnesota Press.

Goriunova, O. (2012), *Art Platforms and Cultural Production on the Internet*, New York and London: Routledge.

Grant, P. (2012), *Imperfection*, Edmonton: AU Press.

Gray, M. L. and S. Suri. (2019), *Ghost Work: How to Stop Silicon Valley from Building a New Global Underclass*, Boston: Houghton Mifflin Harcourt.

Grunenberg, R. (2021), "James Ferraro and Mall Aesthetics," *Ssense*, January 6. Available online: www.ssense.com/en-us/editorial/culture/james-ferraro-and -mall-aesthetics (accessed October 31, 2022).

Guenthner, G. (2016), "Amazon is Crushing Shopping Malls," *Business Insider*, December 20. Available online: www.businessinsider.com/amazon-causing -the-death-of-shopping-malls-2016-12 (accessed October 31, 2022).

Hägglund, M. (2008), *Radical Atheism: Derrida and the Time of Life*, Stanford: Stanford University Press.

Hägglund, M. (2012), *Dying for Time: Proust, Woolf, Nabokov*, Cambridge, MA and London: Harvard University Press.

Hägglund, M. (2019), *This Life: Secular Faith and Spiritual Freedom*, New York: Pantheon Books.

Halter, E. (2009), "The Matter of Electronics," in D. Quaranta (ed.), *Playlist: Playing Games, Music, Art*, 70–7, Gijón: LABoral Centro de Arte y Creación Industrial.

Hansen, M. B. N. (2015), *Feed-Forward: On the Future of Twenty-First-Century Media*, Chicago and London: The University of Chicago Press.

Haraway, D. J. (2016), *Staying with the Trouble: Making Kin in the Chthulucene*, Durham and London: Duke University Press.

Harper, A. (2009), "Hauntology: The Past Inside the Present," *Rouges Foam*, October 27. Available online: rougesfoam.blogspot.com/2009/10/hauntology -past-inside-present.html (accessed October 31, 2022).

Harper, A. (2012), "Comment: Vaporwave and the Pop-Art of the Virtual Plaza," *Dummy*, December 7. Available online: www.dummymag.com/news/adam -harper-vaporwave/ (accessed October 31, 2022).

Hartshorne, C. (1962), *The Logic of Perfection and Other Essays in Neoclassical Metaphysics*, Lasalle: Open Court.

Hayles, N. K. (2017), *Unthought: The Power of the Cognitive Nonconscious*, Chicago and London: The University of Chicago Press.

Heidegger, M. ([1927] 2010), *Being and Time*, trans. J. Stambaugh, revised by D. J. Schmidt, Albany: State University of New York Press.

Helm, S., S. H. Kim, and S. van Riper (2020), "Navigating the 'Retail Apocalypse:' A Framework of Consumer Evaluations of the New Retail Landscape," *Journal of Retailing and Consumer Services*, 54 (3): 101683.

Helmond, A. (2015), "The Platformization of the Web: Making Web Data Platform Ready," *Social Media + Society*, 1 (2): 1–11.

Hertz, G. and J. Parikka (2015), "Zombie Media: Circuit Bending Media Archaeology into an Art Method," in Parikka, *A Geology of Media*, 141–53, Minneapolis and London: University of Minnesota Press.

Hogan, M. (2015), "Data Flows and Water Woes: The Utah Data Center," *Big Data & Society*, 2 (2): 1–12.

Howard, V. (2015), *From Main Street to Mall: The Rise and Fall of the American Department Store*, Philadelphia: University of Pennsylvania Press.

Hughes, F. (2014), *The Architecture of Error: Matter, Measure, and the Misadventures of Precision*. Cambridge, MA and London: MIT Press.

Hui, Y. (2015), "Algorithmic Catastrophe – The Revenge of Contingency," *Parrhesia*, 23: 122–43.

Hussey, A. (2016), "Jennifer O'Connor – Surface Noise," *Pitchfork*, March 7. Available online: pitchfork.com/reviews/albums/21648-surface-noise/ (accessed October 31, 2022).

Ihde, D. (1990), *Technology and the Lifeworld: From Garden to Earth*, Bloomington: Indiana University Press.

Infinity Frequencies (2013), *Computer Death*.

Infinity Frequencies (2014a), *Computer Afterlife*.

Infinity Frequencies (2014b), *Computer Decay*.

Irisarri, R. A. (2019), *Solastalgia*, Room40.

Jackson, D. C. (2019), "Repetition, Feedback, and Temporality in Two Compositions by William Basinski," *Intermédialités/Intermediality*, 33. https://doi.org/10.7202/1065021ar.

Jackson, S. J. (2014), "Rethinking Repair," in T. Gillespie, P. Boczkowski, and K. A. Foot (eds.), *Media Technologies: Essays on Communication, Materiality, and Society*, 221–40, Cambridge, MA and London: MIT Press.

Jameson, F. (1991), *Postmodernism, or, the Cultural Logic of Late Capitalism*, Durham and London: Duke University Press.

JODI (1996–2001), *Untitled Game*.

Jones, P. (2021), *Work without the Worker: Labour in the Age of Platform Capitalism*, London and New York: Verso.

Jucan, B. I., J. Parikka, and R. Schneider (2019), *Remain*, Minneapolis and London: University of Minnesota Press and Meson Press.

Junte, J. (2011), *Hands On: Dutch Design in the 21st Century*, Amsterdam: WBOOKS.

Jury's Out (2017), "Internet Subculture: V a p o r w a v e," *Jury's Out Blog*, Gujarat National Law University, Augustus 4. Available online: jurysoutbl og.wordpress.com/2017/08/04/internet-subculture-v-a-p-o-r-w-a-v-e/ (accessed October 31, 2022).

Kane, C. L. (2019), *High-tech Trash: Glitch, Noise and Aesthetic Failure*, Berkeley: University of California Press.

Kelly, C. (2009), *Cracked Media: The Sound of Malfunction*, Cambridge and London: MIT Press.

Kelly, C., J. Kemper, and E. Rutten, eds. (2021), *Imperfections: Studies in Failures, Flaws, and Defects*, New York: Bloomsbury.

Kemper, J. (2019), "(De)compositions: Time and Technology in William Basinski's 'The Disintegration Loops,'" *Intermédialités/Intermediality*, 33. https://doi.org/10.7202/1065020ar.

Kemper, J. (2022), "The Environment and Frictionless Technology: For a New Conceptualization of the *Pharmakon* and the Twenty-First-Century User," *Media Theory*, 6 (2): 55–76.

Kemper, J. (2023), "Glitch, the Post-Digital Aesthetic of Failure and Twenty-First-Century Media," *European Journal of Cultural Studies*, 26 (1): 47–63.

Kemper, J. and S. Jankowski (forthcoming), "Silicon Valley's Frictionless Future: Cultural and Ecological Ramifications of a Design Philosophy of Seamless Automation," in V. Fors, M. Berg, and M. Brodersen (eds.), *The De Gruyter Handbook of Automated Futures*, Berlin: De Gruyter.

Kemper, J. and D. Kolkman (2019), "Transparent to Whom? No Algorithmic Accountability Without a Critical Audience," *Information, Communication & Society*, 22 (14): 2081–96.

Killeen, Patrick. (2018), "Burned Out Myths and Vapour Trails: Vaporwave's Affective Potentials," *Open Cultural Studies*, 2 (1): 626–38.

King, V., B. Gerisch, and H. Rosa (2019), "'Lost In Perfection' – Ideals and Performances," in V. King, B. Gerisch, and H. Rosa (eds.), *Lost in Perfection: Impacts of Optimisation on Culture and Psyche*, 1–10, New York and London: Routledge.

Kirn, P. (2011), "GlitchHiker: A Game That Dies, Slowly, If You Play Badly," *CDM*, July 5. Available online: cdm.link/2011/07/glitchhiker-a-game-that -dies-slowly-if-you-play-badly (accessed October 31, 2022).

Kittler, F. (2013), *The Truth of the Technological World: Essays on the Genealogy of Presence*, trans. E. Butler, Stanford: Stanford University Press.

Klibansky, R., E. Panofsky, and F. Saxl ([1964] 2019), *Saturn and Melancholia: Studies in the History of Natural Philosophy, Religion and Art*, eds. P. Despoix and G. Leroux, Montreal: McGill-Queen's University Press.

Krapp, P. (2011), *Noise Channels: Glitch and Error in Digital Culture*, Minneapolis and London: University of Minnesota Press.

Krasznahorkai, L. (2016), *War and War*, trans. G. Szirtes, London: Tuskar Rock Press.

Kromhout, M. J. (2021), "Electronic Contingencies: Goeyvaerts' Sine Wave Music and the Ideal of Perfect Sound," in C. Kelly, J. Kemper, and E. Rutten (eds.), *Imperfections: Studies in Failures, Flaws, and Defects*, 129–48, New York: Bloomsbury.

Latour, B. (2009), "A Cautious Prometheus? A Few Steps Toward a Philosophy of Design (With Special Attention to Peter Sloterdijk)," in F. Hackney, J. Glynne and V. Minton (eds.), *Proceedings of the 2008 Annual International Conference of the Design History Society*, 2–10, Florida: Universal Publishers.

Laugier, S. (2015), "The Ethics of Care as a Politics of the Ordinary," *New Literary History*, 46 (2): 217–40.

Laugier, S. (2020), "War on Care," *Ethics of Care*, May 11. Available online: ethicsofcare.org/war-on-care/ (accessed October 31, 2022).

Lefebvre, H. (2014), *Critique of Everyday Life*, trans. J. Moore and G. Elliott, London and New York: Verso.

Lemieux, C. and D. McDonald (2020), *Frictionless: Why the Future of Everything Will Be Fast, Fluid, and Made Just for You*, New York: HarperCollins.

Lialina, O. and D. Espenschied (2009), *Digital Folklore*, Stuttgart: Merz & Solitude.

Lijster, T. (2018), "The Trash of History," in T. Lijster (ed.), *The Future of the New: Artistic Innovation in Times of Social Acceleration*, 217–34, Amsterdam: Valiz.

Lin, M. (2020), "Daniel Lopatin's *Chuck Person's Eccojams Vol. 1* (2010)," in W. Stockton and D. Gilson (eds.), *The 33⅓ B-sides*, 168–71, New York: Bloomsbury.

Lindsheaven Virtual Plaza (2013), *NTSC Memories*, Ailanthus.

Liu, A. (2004), *The Laws of Cool: Knowledge Work and the Culture of Information*, Chicago and London: The University of Chicago Press.

Lovink, G. (2019), *Sad by Design: On Platform Nihilism*, London: Pluto Press.

Lyons, S. (2018), "Ruin Porn, Capitalism, and the Anthropocene," in S. Lyons (ed.), *Ruin Porn and the Obsession with Decay*, 1–10, New York: Palgrave Macmillan.

MacFarlane, R. (2019), *Underland: A Deep Time Journey*, London: Hamish Hamilton.

Macintosh Plus (2011), *Floral Shoppe*, Beer on the Rug.

Manon, H. S. and D. Temkin (2011), "Notes on Glitch," *World Picture*, Winter. Available online: www.worldpicturejournal.com/WP_6/Manon.html (accessed October 31, 2022).

Marenko, B. (2015), "When Making Becomes Divination: Uncertainty and Contingency in Computational Glitch-Events," *Design Studies*, 41: 110–25.

Martin, A., N. Myers, and A. Viseu (2015), "The Politics of Care in Technoscience," *Social Studies of Science*, 45 (5): 625–41.

Mattern, S. (2018), "Maintenance and Care," *Places Journal*, November 13. Available online: placesjournal.org/article/maintenance-and-care/ (accessed October 31, 2022).

Mbembe, A. (2019), *Necropolitics*, trans. S. Corcoran, Durham and London: Duke University Press.

Menkman, R. (2010), *The Collapse of PAL*.

Menkman, R. (2011a), "Rosa Menkman – Collapse of PAL at Trouw (12th of May 2011)," YouTube, May 14. Available online: www.youtube.com/watch?v =DuDwaQDzOZc (accessed October 31, 2022).

Menkman, R. (2011b), *The Glitch Moment(um)*, Amsterdam: Institute of Network Cultures.

Menkman R. (2012), "The Collapse of PAL," *Rhizome*, May 31. Available online: rhizome.org/art/artbase/artwork/the-collapse-of-pal/ (accessed October 31, 2022).

Menkman, R. (2016a), "Elegy for the Collapse of PAL," in R. Bishop, K. Gansing, J. Parikka, and E. Wilk (eds.), *Across & Beyond: A Transmediale Reader on Post-Digital Practices, Concepts, and Institutions*, 114–21, Berlin: Sternberg Press and Transmediale e.V.

Menkman, R. (2016b), "The Collapse of PAL 2010–2012," YouTube, July 9. Available online: www.youtube.com/watch?v=5-XVkI1z1m8 (accessed October 31, 2022).

Menkman, R. (2018), "The Collapse of PAL (2010)," *Beyond Resolution*, June 2. Available online: beyondresolution.info/Collapse-of-PAL (accessed October 31, 2022).

Microsoft (1999), "Friction Free Software to Provide 'Plumbing' for Next Generation of Business-critical Applications," Microsoft, November 8. Available online: https://news.microsoft.com/1999/11/08/friction-free -software-to-provide-plumbing-for-next-generation-of-business-critical -applications/ (accessed March 27, 2023).

Microsoft (2020), "Reducing Friction throughout the Device Lifecycle at Microsoft," Microsoft, April 30. Available online: https://www.microsoft.com /en-us/insidetrack/reducing-friction-throughout-the-device-lifecycle-at -microsoft (accessed October 31, 2022).

Mitchell, M. (2019), *Artificial Intelligence: A Guide for Thinking Humans*, New York: Farrar, Straus and Giroux.

Mladek, K. and G. Edmondson (2009), "A Politics of Melancholia," in C. Strathausen (ed.), *A Leftist Ontology: Beyond Relativism and Identity Politics*, 208–34, Minneapolis and London: University of Minnesota Press.

Moore, G. (2013), "Adapt and Smile or Die! Stiegler Among the Darwinists," in C. Howells and G. Moore (eds.), *Stiegler and Technics*, 17–33, Edinburgh: Edinburgh University Press.

Moradi, I., A. Scott, J. Gilmore, and C. Murphy, eds. (2009), *Glitch: Designing Imperfection*, New York: Mark Batty Publisher.

Müller, K., and R. Aich (2019), "Indian Post-Digital Aesthetics," *Visual Ethnography*, 8 (2): 155–68.

Munster, A. (2011), "From a Biopolitical 'Will to Life' to a Noopolitical Ethos of Death in the Aesthetics of Digital Code," *Theory, Culture & Society*, 28 (6): 67–90.

Muriel, D. and G. Crawford (2018), *Video Games as Culture: Considering the Role and Importance of Video Games in Contemporary Society*, New York and London: Routledge.

Nabokov, V. (1969), *Ada or Ardor: A Family Chronicle*, New York: McGraw Hill.

Nardelli, M. (2009), "Moving Pictures: Cinema and Its Obsolescence in Contemporary Art," *Journal of Visual Culture*, 8 (3): 243–64.

Nemoianu, V. (2006), *Imperfection and Defeat: The Role of Aesthetic Imagination in Human Society*, Budapest: Central European University Press.

Ngai, S. (2012), *Our Aesthetic Categories*, Cambridge, MA and London: Harvard University Press.

Olivier, M. (2015), "Glitch Gothic," in M. Leeder (ed.), *Cinematic Ghosts: Haunting and Spectrality from Silent Cinema to the Digital Era*, 253–70, New York: Bloomsbury.

Openshaw, J. (2015), *Postdigital Artisans: Craftsmanship with a New Aesthetic in Fashion, Art, Design and Architecture*, Amsterdam: Frame Publishers.

Paasonen, S. (2021), *Dependent, Distracted, Bored: Affective Formations in Networked Media*, Cambridge, MA and London: MIT Press.

Parikka, J. (2012), *What is Media Archaeology?*, Cambridge and Malden: Polity Press.

Parisi, L. (2013), *Contagious Architecture: Computation, Aesthetics, and Space*, Cambridge, MA and London: MIT Press.

Parks L. and N. Starosielski, eds. (2015), *Signal Traffic: Critical Studies of Media Infrastructures*, Urbana, Chicago and Springfield: University of Illinois Press.

Peel, J. (2014), "GlitchHiker: The Game That Was Programmed to Die," *PCGamesN*, October 9. Available online: www.pcgamesn.com/indie/glitchhiker-game-was-programmed-die (accessed October 31, 2022).

Peeren, E. (2014), *The Spectral Metaphor: Living Ghosts and the Agency of Invisibility*, New York: Palgrave Macmillan.

Phillips, R. (2020), "Organize Your Own Temporality: Notes on Self-Determined Temporalities and Radical Futurities," in H. Gunkel and k. lynch (eds.), *We Travel the Space Ways: Black Imagination, Fragments, and Diffractions*, 237–43, New Rockford: Transcript Publishing.

Pilkington, M. (2019), *Retail Therapy: Why the Retail Industry is Broken – And What Can Be Done to Fix It*, New York: Bloomsbury.

Pisters, P. (2021), "'I am a Strange Video Loop': Digital Technologies of the Self in Picture-Perfect Mediations," in C. Kelly, J. Kemper, and E. Rutten (eds.), *Imperfections: Studies in Failures, Flaws, and Defects*, 221–41, New York: Bloomsbury.

Plato (2002), *Phaedrus*, trans. Robin Waterfield, Oxford and New York: Oxford University Press.

Poell, T, D. Nieborg, and B. E. Duffy (2021), *Platforms and Cultural Production*. Cambridge: Polity Press.

Polson, J. (2011), "Harder to Judge Than IGF Pirate Kart? Vlambeer's Unplayable GlitchHiker [Interview]," *DIYGamer*, October 30. Available online: web.archive.org/web/20140316042314/http://www.diygamer.com/2011/10/vlambeers-unplayable-playable-igf-entry-glitchhiker-interview/ (accessed October 31, 2022).

Prasad, E. S. (2021), *The Future of Money: How the Digital Revolution Is Transforming Currencies and Finance*, Cambridge, MA and London: Harvard University Press.

Princeton University (2010), "About WordNet," *WordNet*, Princeton University.

Puig de la Bellacasa, M. (2017), *Matters of Care: Speculative Ethics in More Than Human Worlds*, Minneapolis and London: University of Minnesota Press.

Purple Mountains (2019), "That's Just the Way That I Feel," *Purple Mountains [Audio CD]*, Drag City.

Pynchon, T. (1998), *Mason & Dixon*, London: Vintage.

Radulescu, D. (2021), "Dream in a Suitcase: An Immigrant's Transformative Journeys to Imperfect Homes," in C. Kelly, J. Kemper, and E. Rutten (eds.), *Imperfections: Studies in Failures, Flaws, and Defects*, 251–78, New York: Bloomsbury.

Rasch, M. (2020), *Frictie: Ethiek in Tijden van Dataïsme*. Amsterdam: De Bezige Bij.

Reade, J. (2021), "Keeping it Raw on the 'Gram: Authenticity, Relatability and Digital Intimacy in Fitness Cultures on Instagram," *New Media & Society*, 23 (3): 535–53.

Reynolds, S. (2011), *Retromania: Pop Culture's Addiction to its Own Past*, London: Faber and Faber.

Ribeiro, M. H., R. Ottoni, R. West, et al. (2020), "Auditing Radicalization Pathways on YouTube," in *FAT* 20: Proceedings of the 2020 Conference on Fairness, Accountability, and Transparency*, 131–41.

Richardson, M. (2012), "The Disintegration Loops," *Pitchfork*, November 9. Available online: pitchfork.com/reviews/albums/17064-the-disintegration-loops/ (accessed October 31, 2022).

Rid, T. (2017), *Rise of the Machines: The Lost History of Cybernetics*, Melbourne and London: Scribe Publications.

Riofrancos, T. (2019), *Resource Radicals: From Petro-Nationalism to Post-Extractivism in Ecuador*, Durham and London: Duke University Press.

Roberts, S. T. (2019), *Behind the Screen: Content Moderation in the Shadows of Social Media*, New Haven and London: Yale University Press.

Rombes, N. (2009), *Cinema in the Digital Age*, New York: Columbia University Press.

Rosa, H. (2019), *Resonance: A Sociology of Our Relationship to the World*, trans. J. C. Wagner, Cambridge: Polity Press.

Rosenberger, R. and P. Verbeek, eds. (2015), *Postphenomenological Investigations: Essays on Human-Technology Relations*, London: Lexington Books.

Roso, M. (2013), *The Dutch Games Industry: Facts and Figures*. Utrecht: Taskforce Innovation Utrecht Region.

Rutten, E. (2021), "Affirmative Imperfection Rhetoric and Aesthetics: A Genealogy," in T. Barker and M. Korolkova (eds.), *Miscommunications: Errors, Mistakes, Media*, 23–45, New York: Bloomsbury.

Rutten, E. (forthcoming), *Dreams of Imperfection*.

Rutten, E. and R. de Vos (2023), "Trash, Dirt, Glitch: The Imperfect Turn," *European Journal of Cultural Studies*, 26 (1): 3–13.

Ryle, G. ([1949] 2009), *The Concept of Mind*, New York and London: Routledge.

Sadowski, J. (2020), *Too Smart: How Digital Capitalism is Extracting Data, Controlling Our Lives, and Taking over the World*, Cambridge, MA and London: MIT Press.

Saito, Y. (2017a), *Aesthetics of the Familiar: Everyday Life and World-Making*. Oxford: Oxford University Press.

Saito, Y. (2017b), "The Role of Imperfection in Everyday Aesthetics," *Contemporary Aesthetics*, 15 (1): 1–15.

Sandtimer (2015), *Vaporwave is Dead*, Dream Catalogue.

Sarikaya, R. (2018), "Making Alexa More Friction-Free," Amazon Science, April 25. Available online: https://www.amazon.science/blog/making-alexa-more -friction-free (accessed October 31, 2022).

Scharoun, L. (2012), *America at the Mall: The Cultural Role of a Retail Utopia*, Jefferson: McFarland.

Schuppli, S. (2020), *Material Witness: Media, Forensics, Evidence*, Cambridge, MA and London: MIT Press.

Sconce, J. (2000), *Haunted Media: Electronic Presence from Telegraphy to Television*, Durham and London: Duke University Press.

Shayon, S. (2011), "Facebook Unveils Timeline for 'Frictionless' Serendipity," *Brandchannel*, September 22. Available online: web.archive.org/web /20210212013121/www.brandchannel.com/2011/09/22/facebook-unveils -timeline-for-friction-less-serendipity/ (accessed October 31, 2022).

Sobchack, V. (2011), "Media Archaeology and Re-Presencing the Past," in E. Huhtamo and J. Parikka (eds.), *Media Archaeology: Approaches, Applications, and Implications*, 323–33, Berkeley: University of California Press.

Sontag, S. (1981), *Under the Sign of Saturn*, London: Vintage Books.

Srnicek, N. (2017), *Platform Capitalism*, Cambridge: Polity Press.

Star, S. L. (1999), "The Ethnography of Infrastructure," *American Behavioral Scientist*, 43 (3): 377–91.

Starosielski, N. (2015), *The Undersea Network*, Durham and London: Duke University Press.

Sterne, J. (1997), "Sounds Like the Mall of America: Programmed Music and the Architectonics of Commercial Space," *Ethnomusicology*, 41 (1): 22–50.

Stiegler, B. (1998), *Technics and Time, 1: The Fault of Epimetheus*, trans. R. Beardsworth and G. Collins, Stanford: Stanford University Press.

Stiegler, B. (2009a), *Acting Out*, trans. D. Barison, D. Ross, and P. Crogan, Stanford: Stanford University Press.

Stiegler, B. (2009b), *Technics and Time, 2: Disorientation*, trans. S. Barker, Stanford: Stanford University Press.

Stiegler, B. (2011a), *Technics and Time, 3: Cinematic Time and the Question of Malaise*, trans. S. Barker, Stanford: Stanford University Press.

Stiegler, B. (2011b), *The Decadence of Industrial Democracies*, trans. Daniel Ross and Suzanne Arnold, Cambridge: Polity Press.

Stiegler, B. (2013), *What Makes Life Worth Living: On Pharmacology*, trans. Daniel Ross, Cambridge: Polity Press.

Stiegler, B. (2014), *The Lost Spirit of Capitalism*, trans. D. Ross, Cambridge: Polity Press.

Stiegler, B. (2017a), "General Ecology, Economy, and Organology," in E. Hörl and J. Burton (eds.), *General Ecology: The New Ecological Paradigm*, 129–50, New York: Bloomsbury.

Stiegler, B. (2017b), "What Is Called Caring? Beyond the Anthropocene," *Techné*, 21 (2–3): 386–404.

Stiegler, B. (2019), *The Age of Disruption: Technology and Madness in Computational Capitalism*, trans. Daniel Ross, Cambridge: Polity Press.

Strachan, R. (2017), *Sonic Technologies. Popular Music, Digital Culture and the Creative Process*, New York: Bloomsbury.

Syse, K. L. and M. L. Mueller, eds. (2015), *Sustainable Consumption and the Good Life: Interdisciplinary Perspectives*, New York and London: Routledge.

Tanner, G. (2016), *Babbling Corpse: Vaporwave and the Commodification of Ghosts*, Winchester and Washington: Zero Books.

Thomas, M. S. (2008), *Dutch Design: A History*, London: Reaktion Books.

Ticktin, M. I. (2011), *Casualties of Care: Immigration and the Politics of Humanitarianism in France*, Berkeley: University of California Press.

Tiny Mix Tapes (2019), "2010s: Favorite 100 Music Releases of the Decade," December 19. Available online: www.tinymixtapes.com/features/2010s -favorite-100-music-releases-decade (accessed October 31, 2022).

Tolentino, J. (2019), *Trick Mirror: Reflections on Self-Delusion*, London: 4th Estate.

Tsing, A. L. (2005), *Friction: An Ethnography of Global Connection*, Princeton and Oxford: Princeton University Press.

Tsing, A. L., H. A. Swanson, E. Gan, and N. Bubandt, eds. (2017), *Arts of Living on a Damaged Planet*, Minneapolis and London: University of Minnesota Press.

Tuck, E. and C. Ree (2016), "A Glossary of Haunting," in T. E. Adams, S. H. Jones, and C. Ellis (eds.), *Handbook of Autoethnography*, 639–58, New York and London: Routledge.

Van Dijck, J. (2013), *The Culture of Connectivity: A Critical History of Social Media*, Oxford: Oxford University Press.

Van Dooren, T. (2014), *Flight Ways: Life and Loss at the Edge of Extinction*, New York: Columbia University Press.

Van Doorn, N. (2017), "Platform Labor: On the Gendered and Racialized Exploitation of Low-income Service Work in the 'On-Demand' Economy," *Information, Communication & Society*, 20 (6): 898–914.

Virilio, P. (2006), *Speed and Politics*, trans. Mark Polizzotti, Los Angeles: Semiotext(e).

Vlambeer (2011), *GlitchHiker*.

Vlassenrood, L., ed. (2009), *Tangible Traces: Dutch Architecture and Design in the Making*, Rotterdam: NAi Publishers.

Volmar, A. and K. Stine, eds. (2021) *Media Infrastructures and the Politics of Digital Time*, Amsterdam: Amsterdam University Press.

Weiser, M. (1991), "The Computer for the 21st Century," *Scientific American*, September, 78–89.

Wells, K. (2018), "Detroit Was Always Made of Wheels: Confronting Ruin Porn in its Hometown," in S. Lyons (ed.), *Ruin Porn and the Obsession with Decay*, 13–30, New York: Palgrave Macmillan.

Wiener, A. (2020), *Uncanny Valley: A Memoir*. London: 4th Estate.

Williams, E. C. (2011), *Combined and Uneven Apocalypse*, Winchester and Washington: Zero Books.

Wolfenstein osX (2015), "Vaporwave: A Brief History of a Genre That No One Has Heard Of," YouTube, May 3. Available online: www.youtube.com/watch?v=D88iKeigqd4 (accessed October 31, 2022).

Zuboff, S. (2019), *The Age of Surveillance Capitalism: The Fight for a Human Future at the New Frontier of Power*, London: Profile Books.

Index

Note: Page numbers followed with "n" refer to endnotes.